Wood Composites Symposium 2016

Held at the International Woodworking Fair 2016

Atlanta, Georgia, USA
23 August 2016

ISBN: 978-1-5108-3838-3

Printed from e-media with permission by:

Curran Associates, Inc.
57 Morehouse Lane
Red Hook, NY 12571

Some format issues inherent in the e-media version may also appear in this print version.

Printed by Curran Associates, Inc. (2017)

For permission requests, please contact Washington State University (WSU)
at the address below.

Washington State University (WSU)
c/o: Wanda Terry
P.O. Box 645815
Pullman, WA 99164-5815
USA

Phone: (509) 335-6437

Symposium.info@wsu.edu

Additional copies of this publication are available from:

Curran Associates, Inc.
57 Morehouse Lane
Red Hook, NY 12571 USA
Phone: 845-758-0400
Fax: 845-758-2633
Email: curran@proceedings.com
Web: www.proceedings.com

TABLE OF CONTENTS

What Today's Outsourcing Shop Does Right

Mike Lee
Insourcing Specialists

Outsourcing

- Base Theory
  - TCE, Relational View, Core Competency, Resource Base
- Business Structure
  - Where it can be used
- Applications in Operations
  - Materials
  - Processing
  - Technology

Theory of Outsourcing

- Transaction Cost Economics (TCE)
  - Coase and Coase - 1939
    - Easier to make it, than to figure out how to outsource it.
    - Marginal cost of internalization may exceed outsourcing, causing firm stagnation relative to growth.
    - Outsourcing to marginal partners will result in stagnation relative to growth.
- Relational View
  - Dyer & Singh - 1998
    - Complementary Resource Endowments
      - Cabinotch and KCD
    - Inter-firm Specific Assets
      - Cabinotch and Custom Cabinet Shops

Theory of Outsourcing

- Core Competencies
  - Outsource those things that are not core competencies.
    - Moldings
  - Frees firm to focus on core competencies
    - Integration of disparate technologies
  - Does not teach or enable the market to compete
- Resource Based View
  - Grant (1992)
    - identifies five categories of resources: financial, physical, human, technological and reputation.
    - KCD Software
    - Moldings

Structure of a Business

- Finance
  - Capital
  - Accounting systems
    - Payroll
    - AP / AR / GL
- Sales
  - Independent reps.
  - Full time sales force.
  - Marketing / Branding
- Operations
  - Leadership
  - Environmental
  - OSHA
- Purchasing
  - VMI

Applications in Operations

- Technology
  - KCD wrote their own CAD software – for their shop
    - There was not a product available to outsource to.
    - It defined a new "Core Competency"
  - Thousands now buy KCD for CAD-CAM processing
    - There are several choices in the market.
  - Application at Cabinotch: "Relational View"
    - Integrated Design Library to every KCD customer
    - Cutlist designed and owned by Insourcing Specialist – Cabinotch
    - Customer Designs on their software and programs to our plant

Applications in Operations

- Materials – "Core Competencies"
  - Stile and Rail materials are purchased – already molded
    - Vendor Managed Inventories
- While we will use in excess of 750,000 feet of moldings this year, this is not a Core Competency.
- Our vendor delivers twice a week.
- Our vendor manages our inventories – we never place an order.
- The quality of the product is easily in the top decile.
- We price match to Random Links.
- Vendor makes samples of Pre-Production products at no charge with a very fast turn around time.

Applications in Operations

- Sales – "Relational View"
  - Our customer designs with our KCD library from their computer.
    - Our library is usually far more extensive.
  - He gets an immediate price on the cabinet boxes.
  - He gets an immediate price on correctly sized drawers.
  - We deliver in less than 10 days – often less than 5.
  - He carries no inventory – improved cash flow.
  - He can outsource to Cabinotch for a price similar or less than in-house production.
  - Parts are produced correctly on high quality equipment. – No capital investment.
  - He can increase throughput by a factor of 2.5.
  - Innovative assembly system allows flat finishing and fast assembly.

Applications in Operations

- Maintenance
  - TPM is part of the employee culture
- Routers, Sanders, Compressors, Vacuum Systems and Lift Trucks
  - Are maintained under contract with independent experts in the industry on a quarterly or semi-annual basis.
  - Machines are kept in new condition.
  - Downtime is less than 2%
- We do not have the cost of staff maintenance, tools or spare parts.

The Custom Cabinet Shop

- A focus on materials and processing.
  - Design
  - Costing
  - Doors
  - Boxes
  - Moldings
  - Finishing
  - Installation
- Outsourcing is a tool in your belt – take it out when you need it…

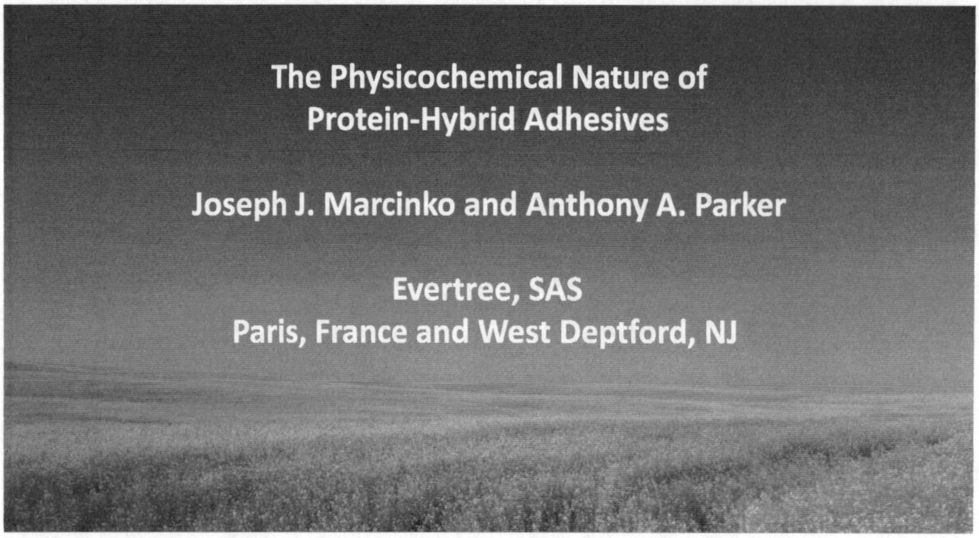

evertree

New Company Overview

evertree

- Avril Group
  €6-billion European leader in oilseeds and proteins contributes investment, R&D resources, feedstock supply chain and industrial capacity
- BPI
  France's public investment bank provides long-term financial backing and industrial relations
- Biopolymer Technologies
  A start-up company with a leading team of American industrial scientists provides the core portfolio of intellectual property and R&D

Avril bpifrance bpt BIOPOLYMER TECHNOLOGIES

evertree http://evertree-technologies.com 2

New Company Overview

- **Our 1st application targets the wood composite panel market:**
  - **Rigorously tested at lab and pilot scale**
  - **Passed multiple, successful industrial trials for end use in North America**

- **Future products made using our technologies will be suitable as**
  - **binders for fiberglass and composites;**
  - **adhesives;**
  - **binders, texturizing agents and dispersants for cosmetics;**
  - **surfactants and reactive components for foams and emulsions**
  - **...**

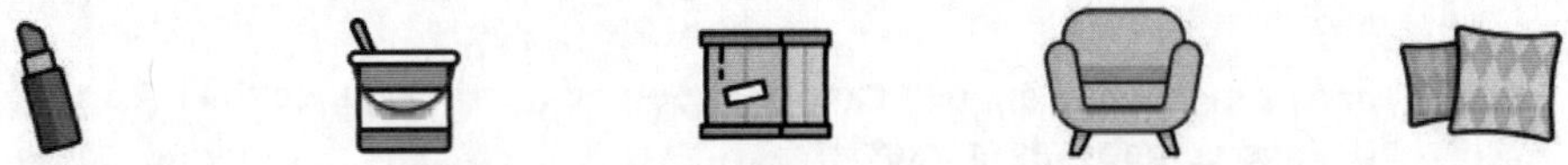

evertree http://evertree-technologies.com 3

The Physicochemical Nature of Protein-Hybrid Adhesives

- **The core technology is based on agricultural meal form many oilseed sources**
- **These agricultural meals share a unique protein structure which allow us to disperse non-reactive oils or reactive oils like PMDI and "protect" the reactive functionality**
- **The use of different meal sources requires slightly different formulation chemistry to compensate for viscosity and oil dispersion differences inherent for each meal**
- **Currently in North America and Europe we are focusing on Canola (Rapeseed) Meal**

Castor Canola Soy Sunflower

evertree

The Physicochemical Nature of Protein-Hybrid Adhesives

- **The main use for Canola/ Rapeseed meal is as animal feed or fertilizer**

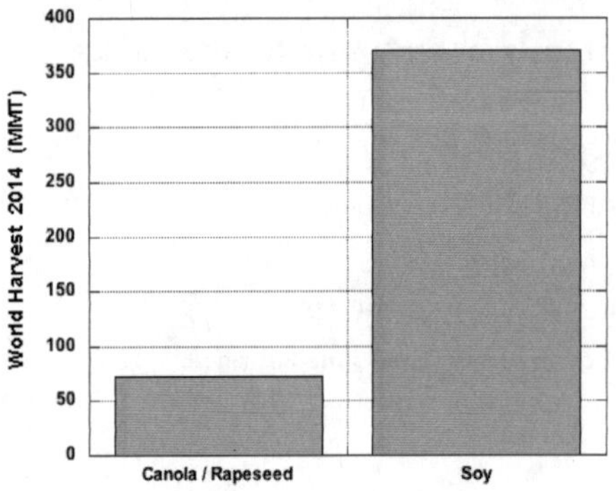

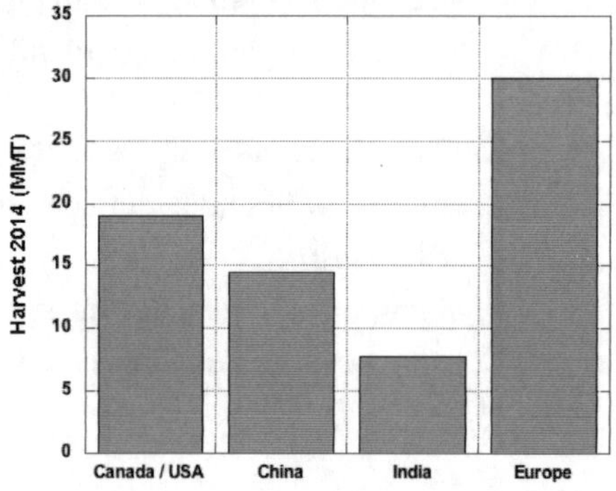

- **There is more than enough Canola/Rapeseed meal to source the NA and EU wood composites market !!!**

evertree

5

The Physicochemical Nature of Protein-Hybrid Adhesives

Processing with nature's help:

- **Rather than following a traditional industrial framework involving multiple steps of chemical processing to design a final product, we try to mimic nature as much as possible, since nature has already given the material a structure that we can use**

- **The chemical potential of the protein fractions are "unlocked" with minimal processing so as to preserve the combined physicochemical and economic benefits**

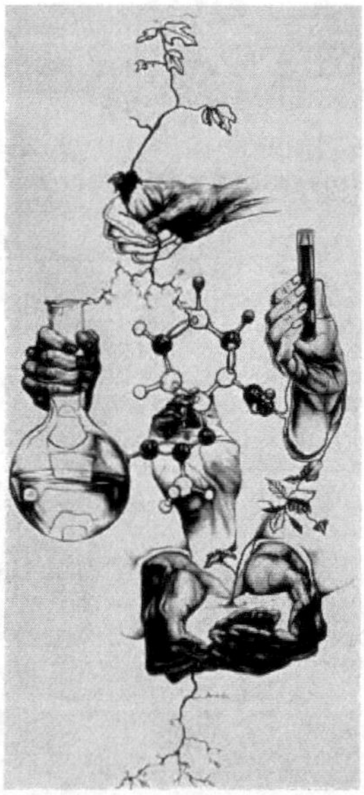

Evertree Wood Composite Adhesives:

- **SynerXiL-WCB® is a ready-to-use, water-based liquid dispersions**

- **SynerXiD-WCB® is available in bulk as a ready-to-mix dry powder**

evertree

6

The Physicochemical Nature of Protein-Hybrid Adhesives

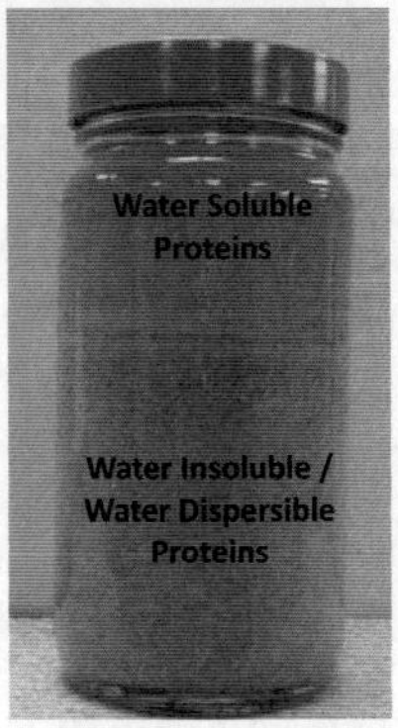

- **Oilseed meals contain proteins (and other chemicals), some which are water soluble and some which are insoluble**
- **There are two fundamental aspects of this chemistry that we use to create adhesive products**
  - **<u>Physical Chemistry</u>:**
    - **The water-insoluble proteins emulsify, disperse and compatibilize dissimilar materials**
    - **We have observed similar chemical characteristics of the water-insoluble proteins from soy, castor, canola, or other oilseeds**
  - **<u>Organic Chemistry</u>:**
    - **We use the available water-soluble protein-amine functionality to create unique reaction products**

evertree

The Physicochemical Nature of Protein-Hybrid Adhesives

The Biology of *Brassica napus* L. (Canola/Rapeseed)

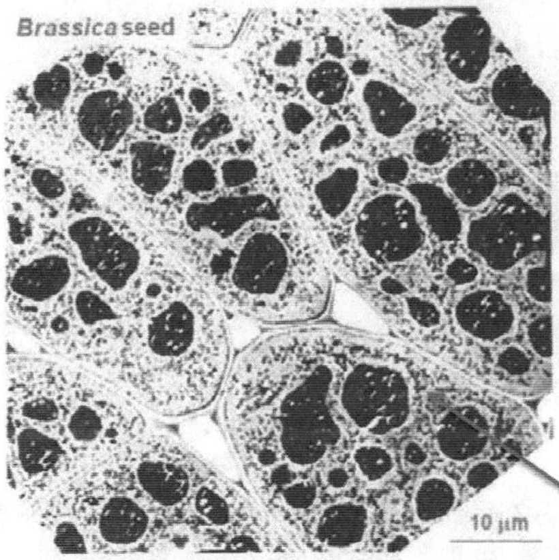

Canola Seed

Representation of an oil body showing the oleosin proteins (yellow), triglycerols (blue), and phospholipds (red)

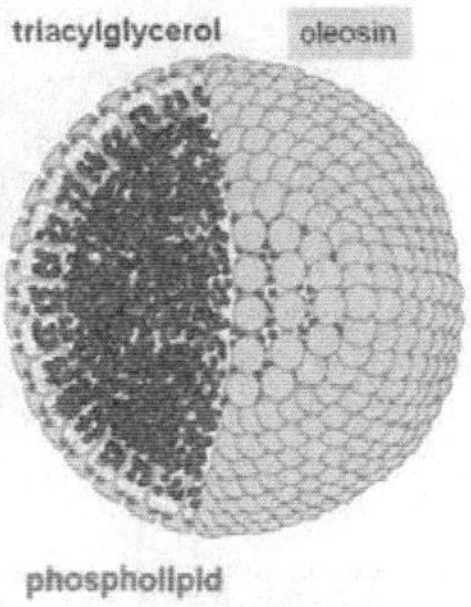

The oil bodies are the small, numerous black dots in the micrograph

evertree

Reference: http://lipidlibrary.aocs.org//plantbio/oilbodies/index.htm
The American Oil Chemists Society, Prof. Anthony H.C. Huang

8

The Physicochemical Nature of Protein-Hybrid Adhesives

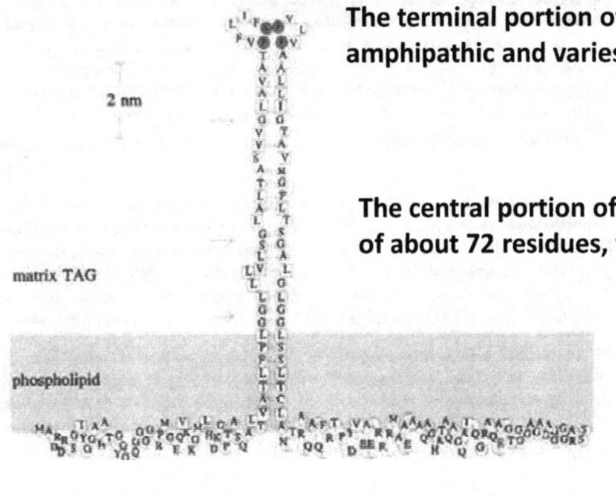

The terminal portion of the protein is hydrophilic or amphipathic and varies in length from about 6 – 68 residues

The central portion of the protein is a long hydrophobic tube of about 72 residues, which only occurs in Oleosin proteins

- **We are investigating, via various analytical techniques, how these proteins and structures interact with reactive molecules like PMDI (or UF and PF) to yield composites that provide high performance and unique physical behavior in their end use applications**

evertree

Reference: http://lipidlibrary.aocs.org//plantbio/oilbodies/index.htm
The American Oil Chemists Society

9

The Physicochemical Nature of Protein-Hybrid Adhesives

- **A peptide or protein chain is essentially a polyamide polymer**

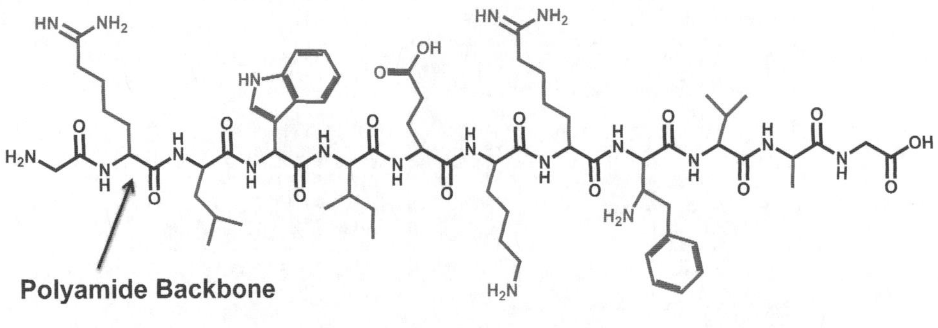

- **The chemical nature of the side chain moieties and their sequencing dictate properties like hydrophobicity, hydrophilicity, radius of gyration, reactivity, and their propensity to associate and assemble**
- **Different oilseeds produce different polyamides with different side-chains and sequences – some are more susceptible to hydrolysis than others**

evertree

The Physicochemical Nature of Protein-Hybrid Adhesives

- **NMR spectra of proteins can be very complicated, so to more easily analyze for chemical structure it is desirable to simplify the analysis in some way**
- **Two Dimensional (2-D) NMR allows you to spread the data out in more than one dimension to help with understanding the configuration of the molecules**

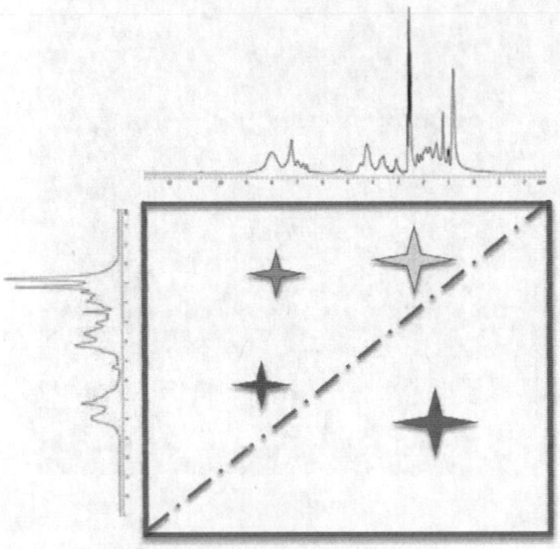

Every 2-D NMR experiment has the same format:

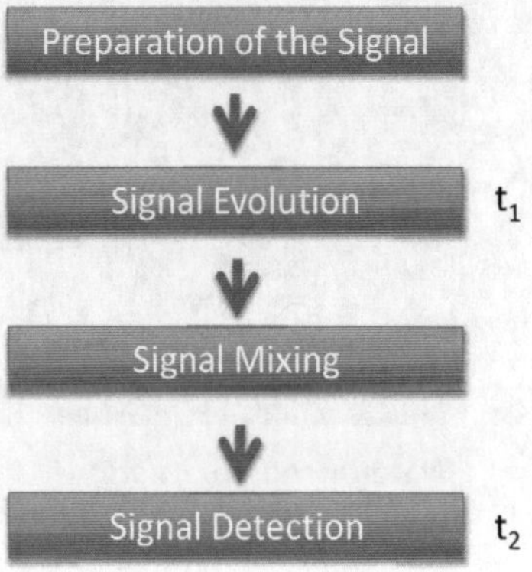

evertree

The Physicochemical Nature of Protein-Hybrid Adhesives

Example of HSQC (one bond couplings) – Protein Meal, which implicitly contains water-soluble and water-insoluble, dispersible fractions (mixture)

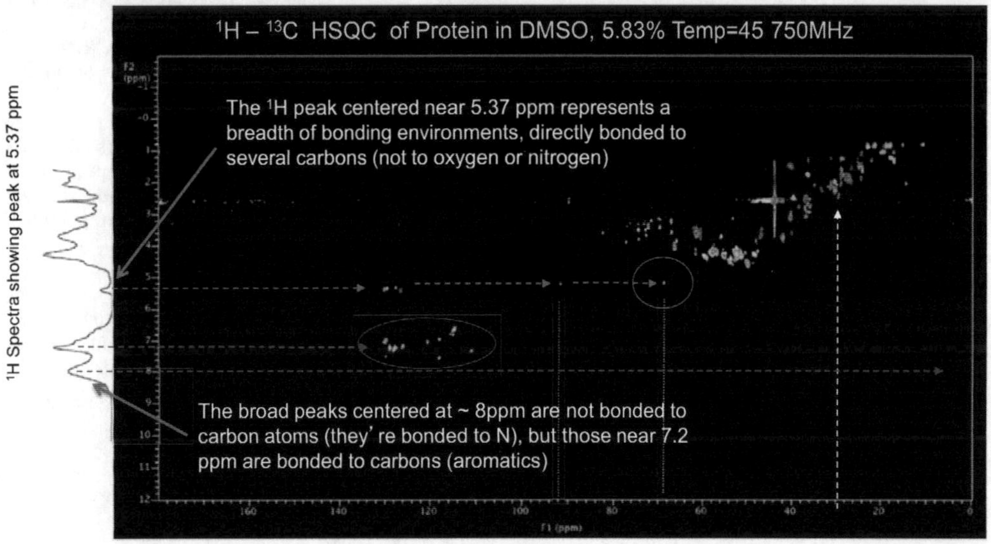

evertree

The Physicochemical Nature of Protein-Hybrid Adhesives

- **The water-insoluble fractions contain fewer carboxylic acids and fewer primary amine functional groups than the water soluble fractions**

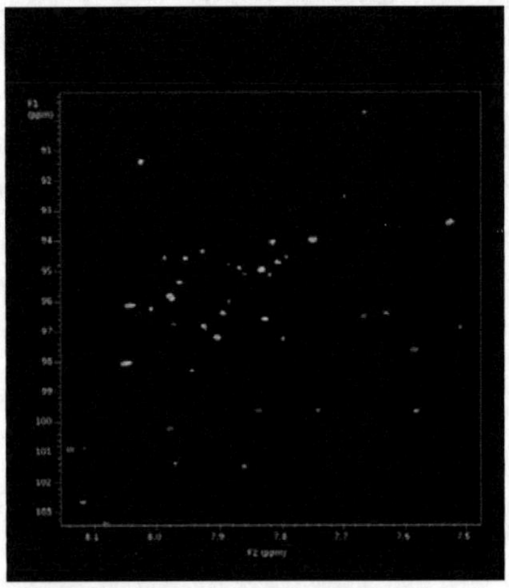

Water insoluble fraction

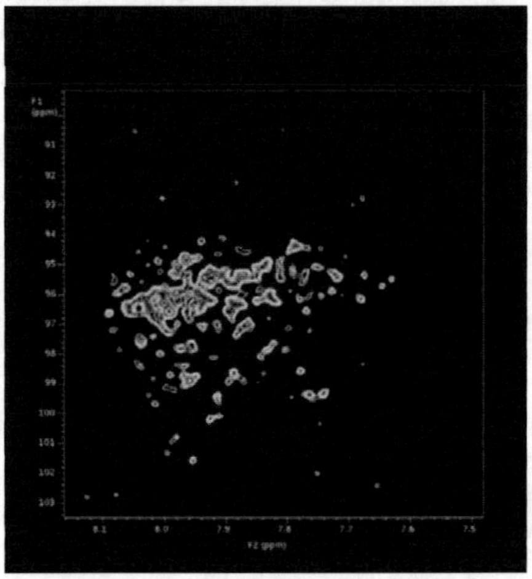

Mixture
(Soluble and Insoluble)

evertree

The Physicochemical Nature of Protein-Hybrid Adhesives

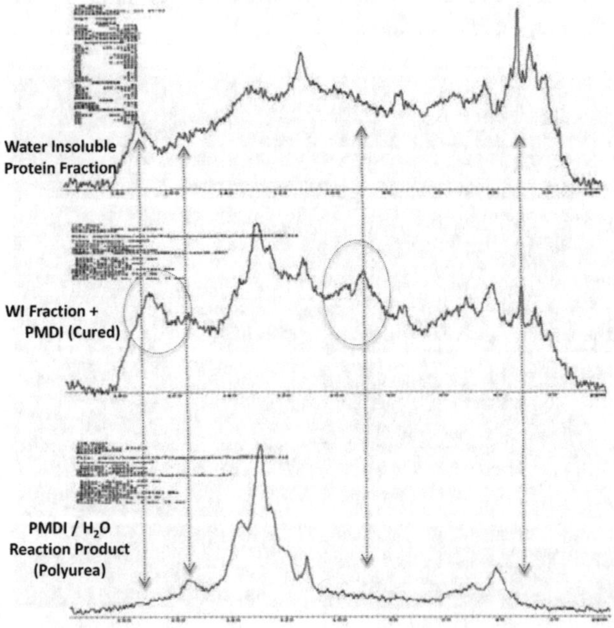

Solid State NMR, Bloch Decay Spectra

- **Raw meal contains both Water Insoluble (WI) and Water Soluble (WS) protein fractions**
- **The greatest changes in the reacted WI Fraction + PMDI sample (center sample) as compared to the control samples are in the region 80 – 100 ppm and the carbonyl region centered at 170 ppm**
- **The changes in the 80 – 100 ppm region may be due to resolution enhancement due to molecular mobility differences in this sample**

The shift in the carbonyl peak position is due to reaction of PMDI with polypeptide moieties

evertree

14

The Physicochemical Nature of Protein-Hybrid Adhesives

Time-Lapsed FTIR Spectra of PMDI/Canola Dispersions

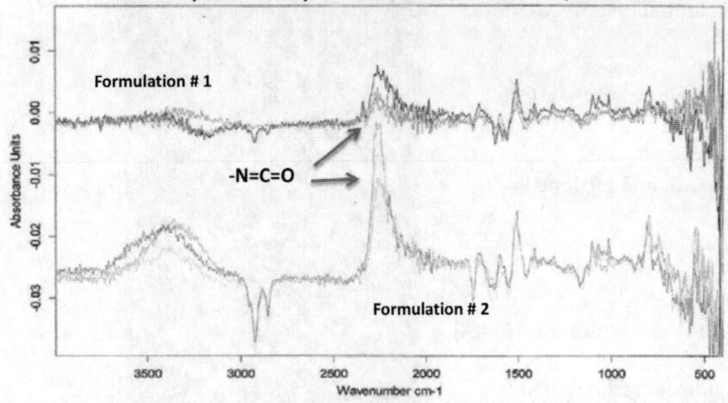

- **SATR-FTIR of PMDI/Canola dispersions vs. time**
- **Spectra were collected at t = 20 minutes, 1 hr., and 2.5 hrs. after mixing**

- **The presence of active NCO moieties has been confirmed by FTIR measurements that show active NCO moieties are present in the dispersions for long periods of time**

evertree

The Physicochemical Nature of Protein-Hybrid Adhesives

- **Why do we study oilseed biology and chemistry?**

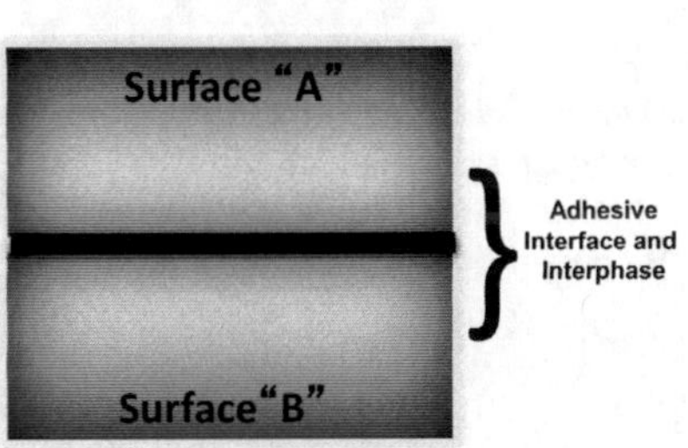

Adhesion requires the interaction of molecules !!

evertree

16

The Physicochemical Nature of Protein-Hybrid Adhesives

- **Evertree's products offer substantial benefits to the Wood Composite Industry:**
  - **Cost-effectively reduce fossil-based chemicals in industrial processes**
  - **Improve occupational health and hygiene in manufacturing facilities**
    - **Reduced PMDI emissions**
    - **Clean-up with H_2O**
  - **Dramatic reduction in press release agents (PMDI)**
  - **Achieve statistically significant reduction in quality assurance variability**
  - **Increase product quality and competitiveness by adding additives and reactants in standard formulations**

evertree

17

Adhesive Preparation for PB, MDF, and OSB Wood Composite

- **SynerXiL-WCB ® product is a water based dispersion (dry content of 20 to 30%), made from Canola meal as main raw-material**
- **The typical ratio of PMDI /Protein is 1:1 to 3 : 1 depending on application and performance criteria**
- **The SynerXiL and PMDI can be mixed either in-line or at the dual port spinning disk (OSB)**
- **Flushing and cleaning of the lines, pumps, and atomizers is done with the Meal/Water/Urea dispersion, followed by a water purge. No solvent is necessary**

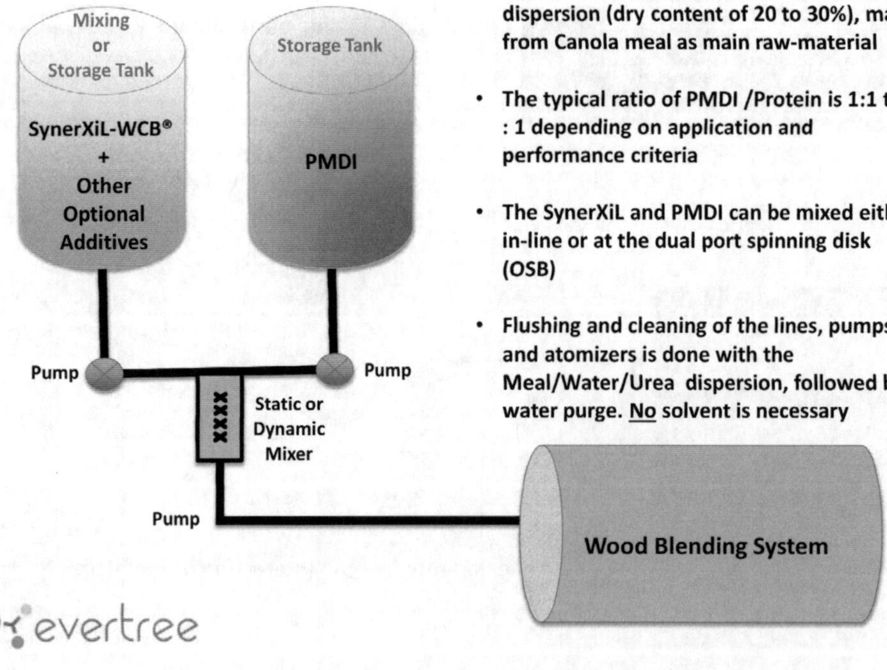

evertree

The Physicochemical Nature of Protein-Hybrid Adhesives

- **The following data was collected at The University of Maine**
- **Boards were made with high loadings of PMDI (specialty boards) and boards were made with low loadings of PMDI (commodity boards)**
- **Flexural Properties (MOR and MOE) were measured along with Thickness Swell (TS) and Internal Bond (IB)**
- **In this study equivalent or better properties were obtained with Evertree formulations that provide a reduction in PMDI loading of approximately 33% to 50 %**
- **These findings have since been confirmed in large scale, production trials**

evertree

The Physicochemical Nature of Protein-Hybrid Adhesives

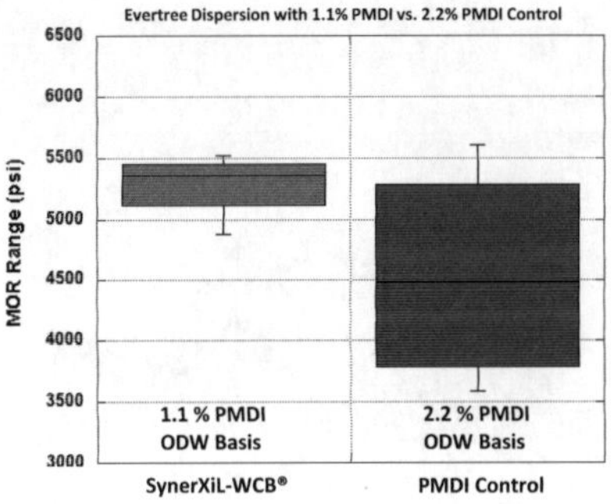

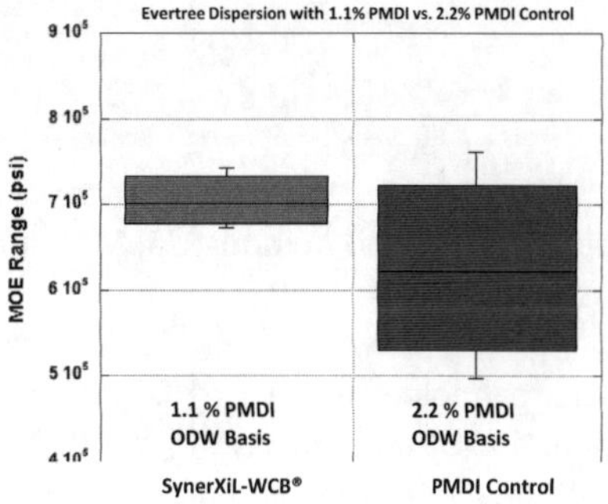

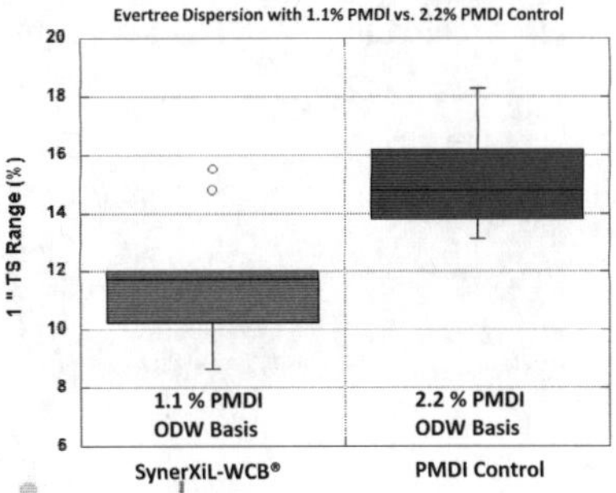

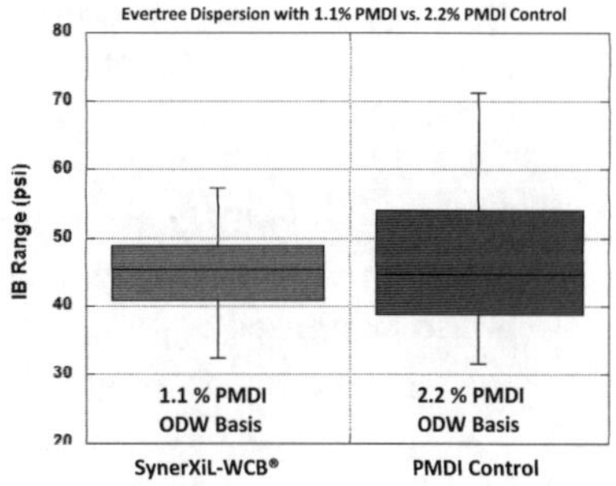

evertree

The Physicochemical Nature of Protein-Hybrid Adhesives

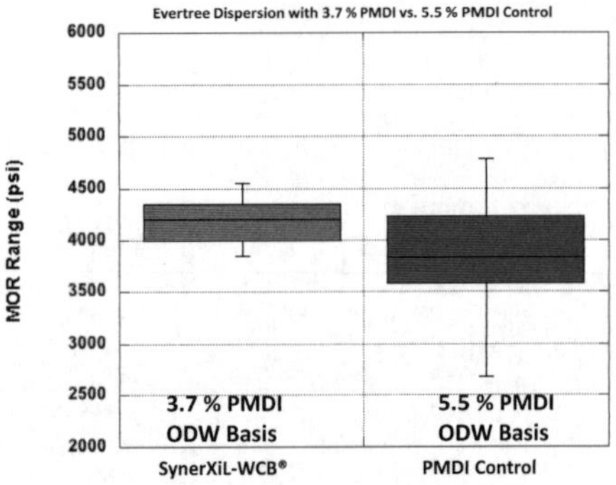

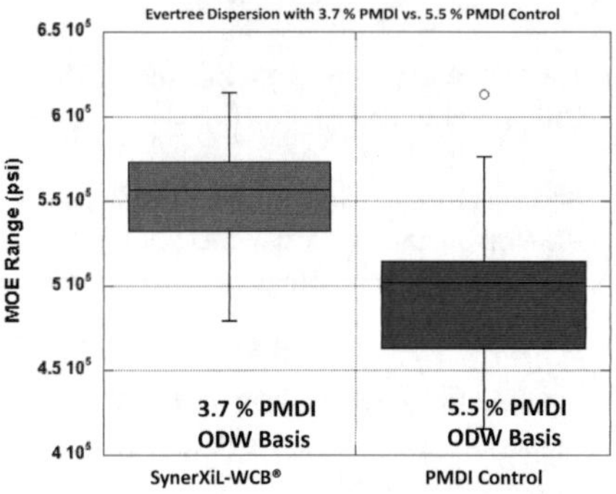

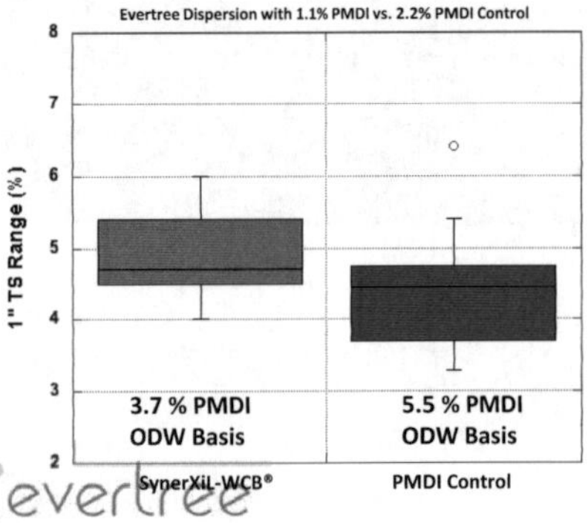

- **We consistently see narrower statistical distributions for the PMDI/Canola Hybrid adhesive**
- **The OSB findings have been reproduced on the industrial scale in an OSB manufacturing plant**

evertree

The Physicochemical Nature of Protein-Hybrid Adhesives

- **The Evertree Protein/Hybrid technology has also been tested in industrial MDF and PB manufacturing plants**
- **The technology in MDF / HDF allows for injection of the adhesive right after the refiner, just as you would inject UF resin**
  - **This allows for very efficient adhesive distribution of the PMDI / Protein Hybrid adhesive on the fibers**
  - **Because the PMDI is "protected" by the proteins in the meal blow-line build-up is also minimized**

evertree

The Physicochemical Nature of Protein-Hybrid Adhesives

HDF Property Comparison, Plant # 1

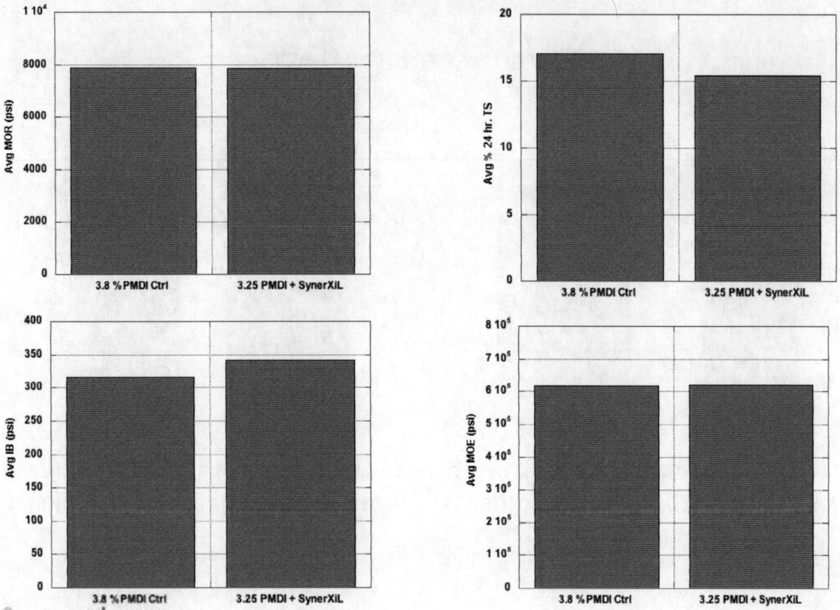

The Physicochemical Nature of Protein-Hybrid Adhesives

HDF Property Comparison, Plant # 2

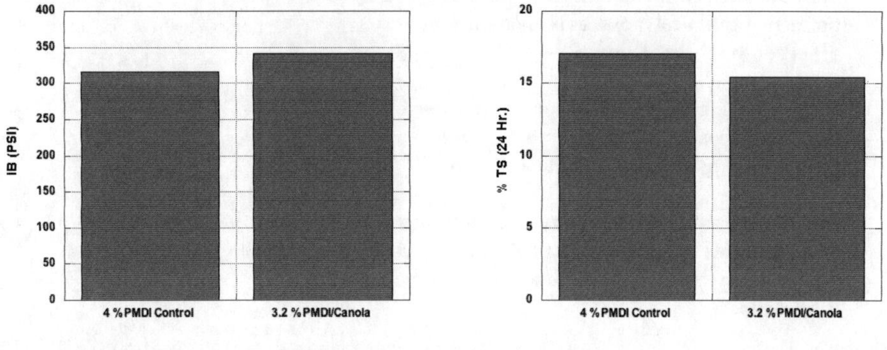

- **The PMDI/Canola boards all met or exceeded plant specifications with a 20 % reduction in PMDI**
- **The plant was able to speed their production rate (continuous press) by 20 %**
- **The plant was able to reduce their external release by 30 %**

The Physicochemical Nature of Protein-Hybrid Adhesives

HDF Property Comparison, Plant # 2

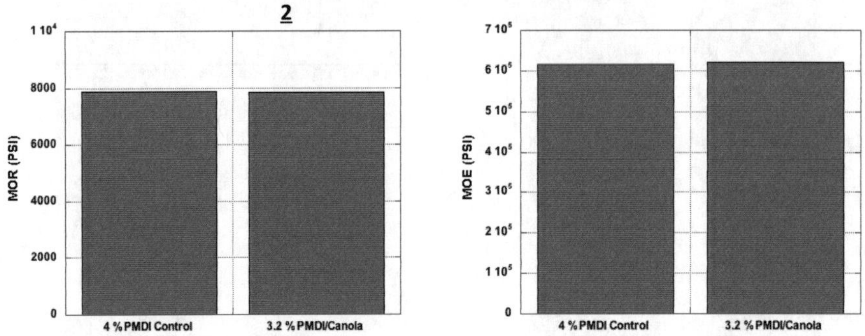

- **The plant was able to get higher face density of the board even with an increase in press speed by 20 %. The implications of this is that they can reduce PMDI loading even more and maintain board stiffness**

evertree

The Physicochemical Nature of Protein-Hybrid Adhesives

- **The fundamental characteristic of dispersing and protecting chemical moieties is an important aspect of the technology**
- **Mixtures of ingredients that would otherwise be incompatible with one another in the absence of the Evertree technology is easily achieved**
- **Another advantage is the ability to incorporate small amounts of additives, like a Si fluid, which would otherwise be impossible to add efficiently to the wood**
- **Evertree is working on an all bio-based product which will match or enhance the performance of the current technology**

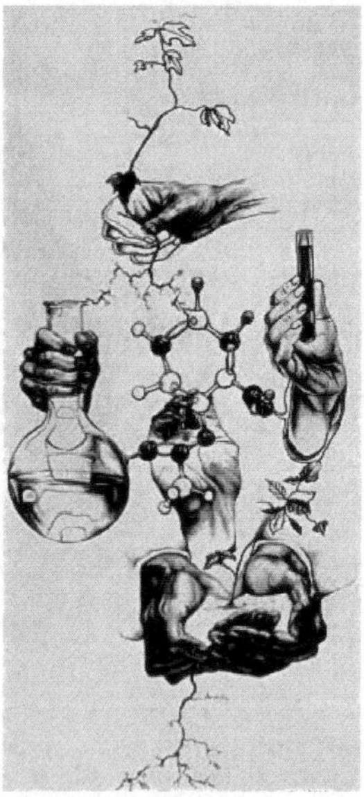

evertree

The Physicochemical Nature of Protein-Hybrid Adhesives

Summary

- **It is important to understand the physicochemical nature of any adhesive**
- **The nature of the molecular interactions between the protein and PMDI are complicated and fundamental research into the interactions is ongoing, but we do have NMR and IR spectroscopic data showing reaction products**
- **Investigations of the hybrid adhesives prepared with UF, PF and MUF are also ongoing**
- **The data allows us to design adhesive technologies to meet specific WC applications and exploit the oilseed meal proteins ability to disperse reactive and non-reactive hydrophobic materials**
- **The "protection" of the reactive -NCO groups by the proteins provides many advantages**
  - **Reduced PMDI consumption**
  - **Reduced blow-line build-up**
  - **Reduced press sticking**
  - **Reduced PMDI emissions (in PB and OSB)**
  - **The ability to mix adhesive systems easily (e.g. SynerXiL © /PMDI/UF)**

evertree

27

"All truths are easy to understand once they are discovered; the point is to discover them."

Galileo Galilei

Astronomer, Physicist, Engineer, Mathematician, and Philosopher

1564 – 1642

evertree

HEXION

A Life In the Timber Industry

A Focus on Stewardship

Safety is #1 Get Zero Get Home

August, 2016

HEXION

A Life In the Timber Industry

What does that mean when I say it; "Thanks for asking", "Perspective"

- **You have one life and one life only to trade for all the decisions you make with your time.**
- **Luckily; the Timber Industry turns out to be my passion; and I have no regrets in total; I do regret a couple minor forays on the journey to what is now 40 years in Industry. The aggregate of what I've traded my only life for is and has been equally wonderful and rewarding.**
- **So when you think about this Industry, it's colorful and interesting history, the players, the innovations, and the seemingly endless evolution of it; ask yourself; "Am I willing to trade my only life to be part of this?"**
- **A short resume; Vanply, Southwest Forest Products, Willamette Industries, Borden Chemical, Hexion Specialty Chemicals, Momentive Specialty Chemicals, and now back to Hexion.**
- **Roles at each; Busheled on the spreaders, and did many of the jobs in the plant, Quality Control Supervisor, Technical Director for Plywood and Sawmill, Technical Manager for Borden, Sales Manager for Borden Eastern US, Sales Manager for Performance Adhesives Western US, and for the last 6 years; N.A. Business Director for Performance Adhesives; one of 4 Business Units making up N.A. Forest Products.**

2

HEXION

A Life In the Timber Industry

• What is Important to me about this life in the industry?

- Learning everything I could to create Subject Matter Expertise. (Value).
- Becoming an exceptional listener; this is a skill you'll want to make a life's work.
- (A brief description of "High Performance Listening"), key to communication.
- A willingness to be mentored; (for you younger folks; a Paramount concept).
- A willingness to be a mentor; (for us more senior folks; equally Important).
- Understanding that you choose to be part of "The Art of Conversion".
    - There really is as much art as science in what we do.
    - Turning "Ideas Into Dollars".
    - Everything between the idea & the dollar has to work, to get the dollar.
- Turning cylinders into rectangles out of sustainable materials while practicing stewardship.

3

HEXION

What does "Stewardship & Sustainability" mean?

Responsible harvesting and Protection are the reasons behind the success story.

Here is a definition from the FAO; (Food and Agriculture Organization).

The definition of <u>sustainable</u> forest management, as understood by the FAO is: The <u>stewardship</u> and use of forests and forest lands in a way, and at a rate, that maintains their biodiversity, productivity, regeneration capacity, vitality and their potential to fulfill, now and in the future, relevant ecological, economic and social functions, at local, national, and global levels, and that does not cause damage to other ecosystems. These rules now govern how forests are managed.

4

HEXION

More trees than there were 100 years ago? It's true!

This is great news for those who care about the environment because trees Sequester & Store CO_2;

They Produce Oxygen — which is necessary for all life on Earth — remove toxins from the air, and create habitat for animals, insects, & basic forms of life.

Well-managed forest plantations like those overseen by the Forest Stewardship Council also furnish us with wood, a renewable material that can be used for building, furniture, paper products and more, and all of which are biodegradable at the end of their lifecycle.

5

HEXION

Evolution & Innovations

Product & Process Examples;

- Plywood.
- OSB; Oriented Strand Board; can be made from thinning operations.
- LVL; Laminated Veneer Lumber.
- Laminated Beams; replaces solid sawn large timbers.
- Rig mats to protect the tundra in sensitive areas.
- LSL; Laminated Strand Lumber; extremely strong; surpasses solid sawn lumber.
- CLT; Cross Laminated Timber for tall wood buildings.

- What's next?
- Ideas; that's what next. Ideation sessions intracompany.
- Pick the best, table the rest for later.

6

HEXION

Stewardship in many forms.

People **Process** **Products**

People:

People are the greatest asset an organization has.

Recruiting and Retaining talent is key to success.

Succession planning is a process you'll not regret.

Process:

Building & refining the process never ends; it is the vehicle you build, maintain, and steer every day; it delivers the value of the ideas you were right about.

Products: They deliver Quality, Consistency & Value. Build things people want and that your people & process are good at making.

7

HEXION

Stewardship in many forms.

Formaldehyde; a molecule folks like to beat up on yet occurs naturally.

More importantly, from a toxicological perspective, formaldehyde is also a natural product of metabolism. According to ATSDR's Toxicological Profile for Formaldehyde, formaldehyde is a normal metabolic product that is present at all times throughout the body and is rapidly metabolized, and, formaldehyde induces adverse effects (predominantly eye, nose and throat irritation in humans) only at the portal of entry, i.e., the upper respiratory tract, in the case of inhalation exposures under 5-10 ppm.

As an essential metabolic intermediate in the single-carbon pool, large amounts of formaldehyde (i.e., more than 50,000 milligrams) are produced, used, and/or detoxified by the human body every day.

Nevertheless, exposures are so much lower by comparison that even relatively high exposures by either inhalation in the workplace (up to 15 ppm), or ingestion of foods naturally high in formaldehyde (e.g., smoked meats, spinach, pears, and provolone cheese), have no detectable impact on normal blood levels of endogenous formaldehyde.

Formaldehyde is excreted in urine as formic acid, and in exhaled breath as either CO_2, or unchanged formaldehyde. Median concentrations of formaldehyde in exhaled human breath are 4-10 ppb [26,27] and individual samples in one study ranged from 1.23 to 72.73 ppb [26]. Thus, humans normally exhale formaldehyde concentrations roughly a thousand times higher than the 1-in-a-million risk level (0.01 ppb in air) that would correspond to EPA's recommended new Unit Risk for Formaldehyde (1×10^{-4} $(ppb)^{-1}$). [14]

Author: Frank Schnell

Reference: Formaldehyde: Fact & Fiction

8

HEXION

Stewardship in many forms.

What is U.S. electricity generation by energy source?

In 2015, the United States generated about 4 trillion kilowatt hours of electricity.1 About 67% of the electricity generated was from fossil fuels (coal, natural gas, and petroleum).

Major energy sources and percent share of total U.S. electricity generation in 2015:1

- **Coal = 33%**
- **Natural gas = 33%**
- **Nuclear = 20%**
- **Hydropower = 6%**
- **Other renewables = 7% • Biomass = 1.6%**
- **Geothermal = 0.4%**
- **Solar = 0.6%**
- **Wind = 4.7%**

- **Petroleum = 1%**
- **Other gases = <1%**

Ask me where Gypsum comes from?

9

HEXION

Stewardship in many forms.

Coal burning electric plants move fly ash into wallboard manufacture?

In 1999, about 2.8 Mt of synthetic gypsum produced as FGD material by electric utilities went into wallboard manufacture.

The synthetic gypsum meets and often exceeds the specifications for wallboard manufacture. Some wallboard plants that will use 100 percent synthetic gypsum are being built and some have started production (Drake, 1997; Olson, 2000). In 1999, synthetic gypsum accounted for about 17 percent of the total gypsum used in wallboard manufacture, and this figure is expected to increase.

“Be discerning in the causes you take up, and the positions you believe; check the facts; Stewardship is the key, to simply eliminate something versus studying the potential unintended consequences can be worse than the perception of the cause itself.”

10

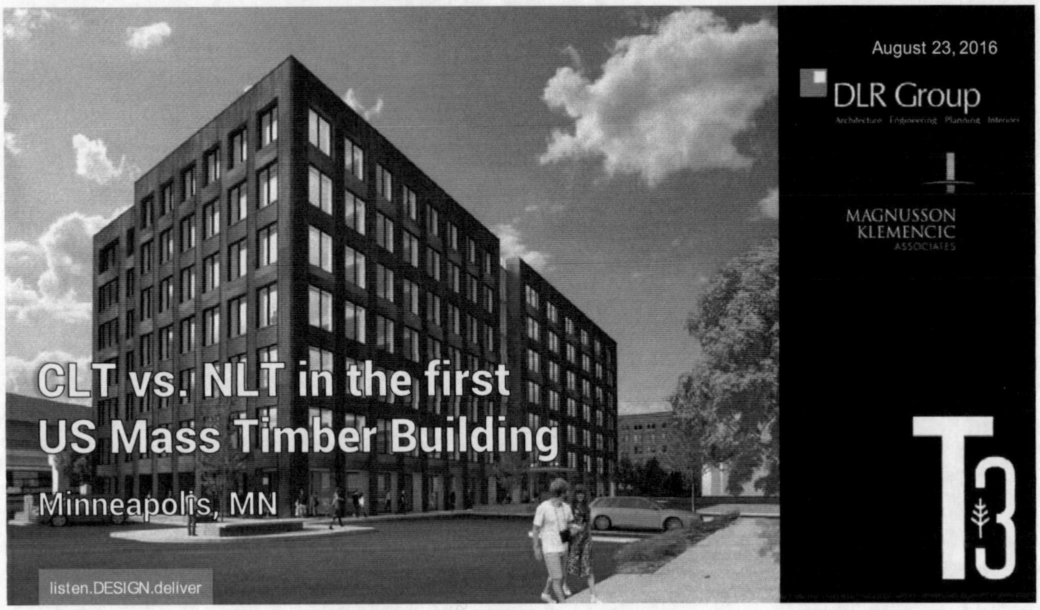

Elevate the Human Experience through Design

1 TEAM

2 INTRODUCTION

3 DESIGN

4 MATERIALS

Team

Hines Interests Limited Partnership- Client
Michael Green Architecture- Design Architect
DLR Group- Architect of Record
Magnusson Klemencic Associates- Structural Engineer
Dunham- Mechanical, Electrical, Plumbing

Leif Johnson

Magnusson Klemencic Associates
P.E., S.E., LEED AP / Senior Associate

Amado Guevara III

DLR Group
Senior Architect / Associate

DLR Group | Magnusson Klemencic Associates

Introduction

T3- Timber, Transit + Technology

- First multi-story commercial HT building built in the US in the last 100 years
- 7 story office building + 1 lower level of parking
- 29,000 SF concrete pedestal main level
- Roughly 198,000 SF of timber framing on levels 2-7

DLR Group | Magnusson Klemencic Associates

The Old / New Office

Pros:
- Open spaces
- Warmth
- Character

Cons:
- Drafty
- Creaky floors
- Lack of integration of technology
- Small exterior windows

DLR Group | Magnusson Klemencic Associates

DLR Group | Magnusson Klemencic Associates

The T3 Solution

Pros:
- Open spaces
- Warmth
- Character
- Sound insulated floors
- Integrated technology
- Shallow beams at core "hide" the mechanical systems
- Large windows

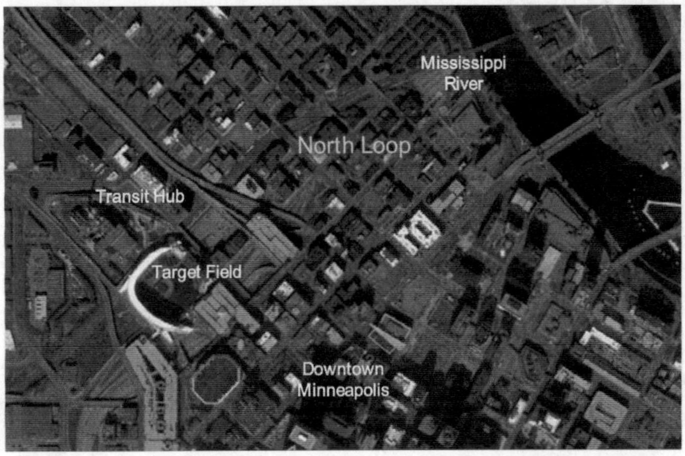

DLR Group | Magnusson Klemencic Associates

Urban Location

- Urban Core amenities
- Mass Transit + Walkability

T3 Site

- North Loop Neighborhood
- Commuter friendly
- Bike trail integration

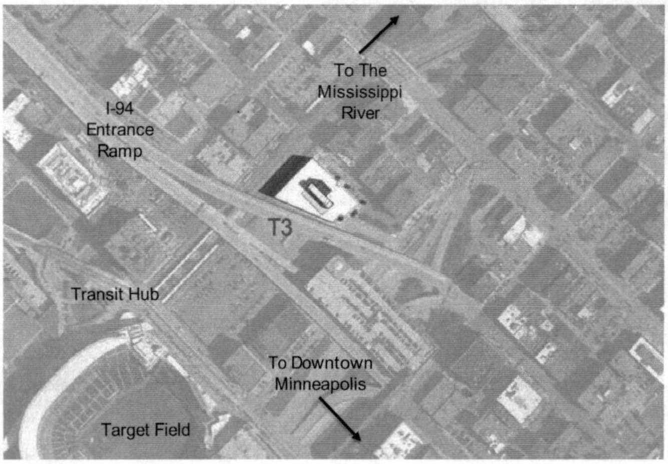

DLR Group | Magnusson Klemencic Associates

Design

DLR Group | Magnusson Klemencic Associates

Ground Level

Tenant Amenities:

- Lobby- entry to retail
- Social Workspace- small-group meeting areas
- Bike Repair Area
- Bike Storage
- Locker Rooms

T3

DLR Group | Magnusson Klemencic Associates

Ground Level

Tenant Amenities:

- Lobby- entry to retail
- Social Workspace- small-group meeting areas
- Bike Repair Area
- Bike Storage
- Locker Rooms

T3

Roof Level

Tenant Amenities:

- Rooftop patio with tables and seats, gas grills and views of downtown Minneapolis.

DLR Group | Magnusson Klemencic Associates

Materials

DLR Group | Magnusson Klemencic Associates

Exterior

Materials:
- Pre-weathered steel panels
    - Flat
    - Corrugated
- Black aluminum panels
- Board-formed concrete
- Glass

DLR Group | Magnusson Klemencic Associates

Exterior

Materials:
- Pre-weathered steel panels
    - Flat
    - Corrugated
- Black aluminum panels
- Board-formed concrete
- Glass

DLR Group | Magnusson Klemencic Associates

Exterior

Materials:
- Pre-weathered steel panels
  - Flat
  - Corrugated
- Black aluminum panels
- Board-formed concrete
- Glass

T3

DLR Group | Magnusson Klemencic Associates

Interior

Materials:
- Glulam beams and columns
- NLT floor/ceiling system
- Pre-weathered steel panels
- Blackened steel panels
- Board-formed concrete

T3

DLR Group | Magnusson Klemencic Associates

Interior

Materials:
- Glulam beams and columns
- NLT floor/ceiling system
- Pre-weathered steel panels
- Blackened steel panels
- Board-formed concrete

T3

DLR Group | Magnusson Klemencic Associates

Interior

Materials:
- Glulam beams and columns
- NLT floor/ceiling system
- Pre-weathered steel panels
- Blackened steel panels
- Board-formed concrete

T3

Interior

Materials:
- Glulam beams and columns
- NLT floor/ceiling system
- Pre-weathered steel panels
- Blackened steel panels
- Board-formed concrete

DLR Group | Magnusson Klemencic Associates

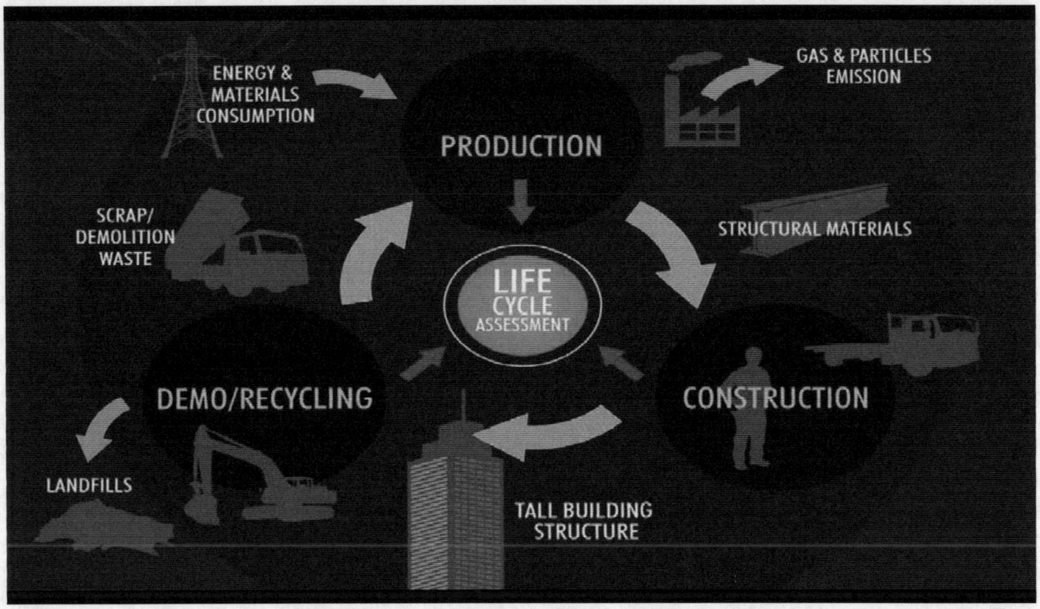
ENERGY &
MATERIALS
CONSUMPTION
PRODUCTION
GAS & PARTICLES
EMISSION
SCRAP/
DEMOLITION
WASTE
STRUCTURAL MATERIALS
LIFE
CYCLE
ASSESSMENT
DEMO/RECYCLING
CONSTRUCTION
LANDFILLS
TALL BUILDING
STRUCTURE

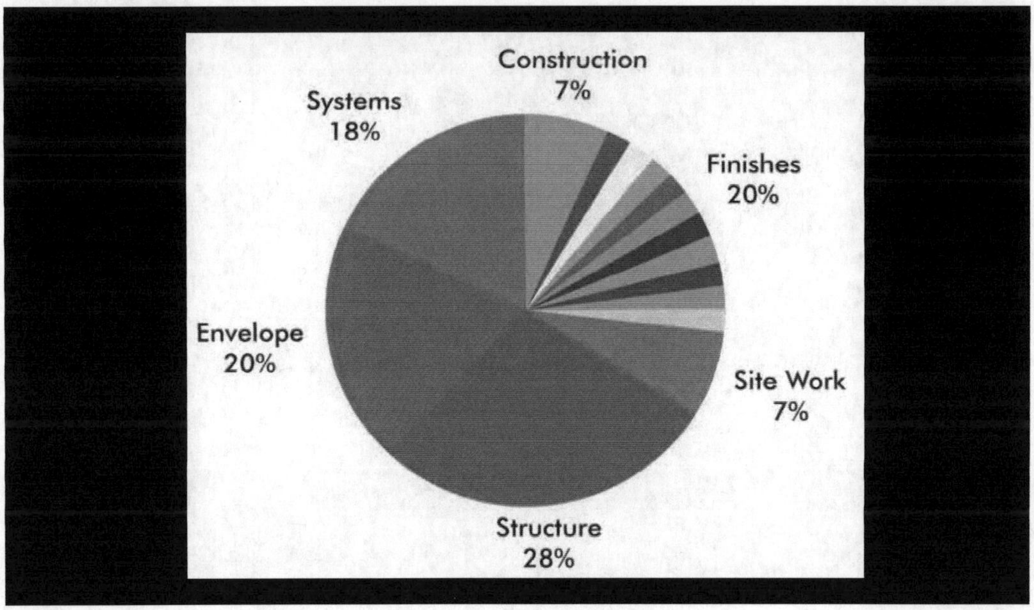
Construction
7%
Systems
18%
Finishes
20%
Envelope
20%
Site Work
7%
Structure
28%

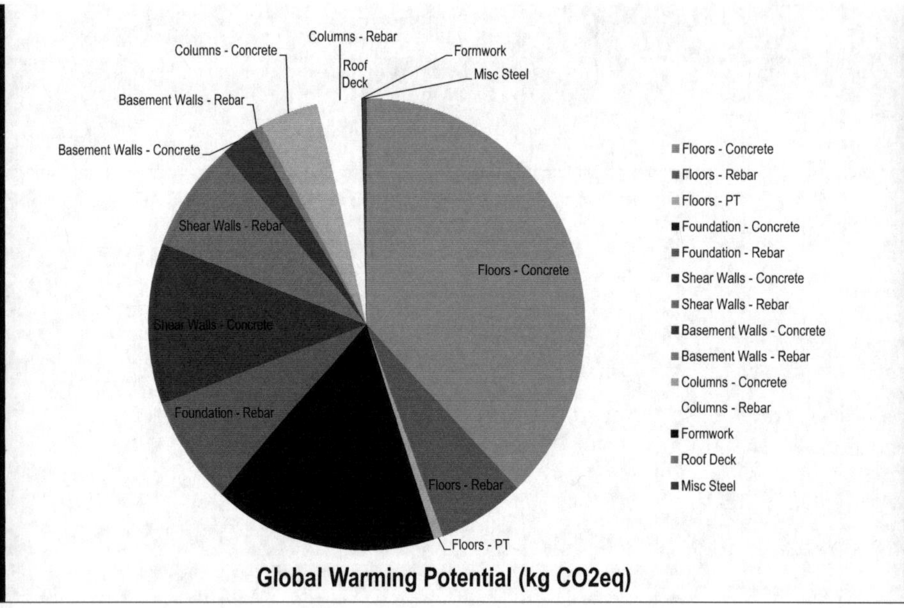
Columns - Rebar
Columns - Concrete
Roof Deck
Formwork
Misc Steel
Basement Walls - Rebar
Basement Walls - Concrete
Shear Walls - Rebar
Floors - Concrete
Shear Walls - Concrete
Foundation - Rebar
Floors - Rebar
Floors - PT
Floors - Concrete
Floors - Rebar
Floors - PT
Foundation - Concrete
Foundation - Rebar
Shear Walls - Concrete
Shear Walls - Rebar
Basement Walls - Concrete
Basement Walls - Rebar
Columns - Concrete
Columns - Rebar
Formwork
Roof Deck
Misc Steel
Global Warming Potential (kg CO2eq)

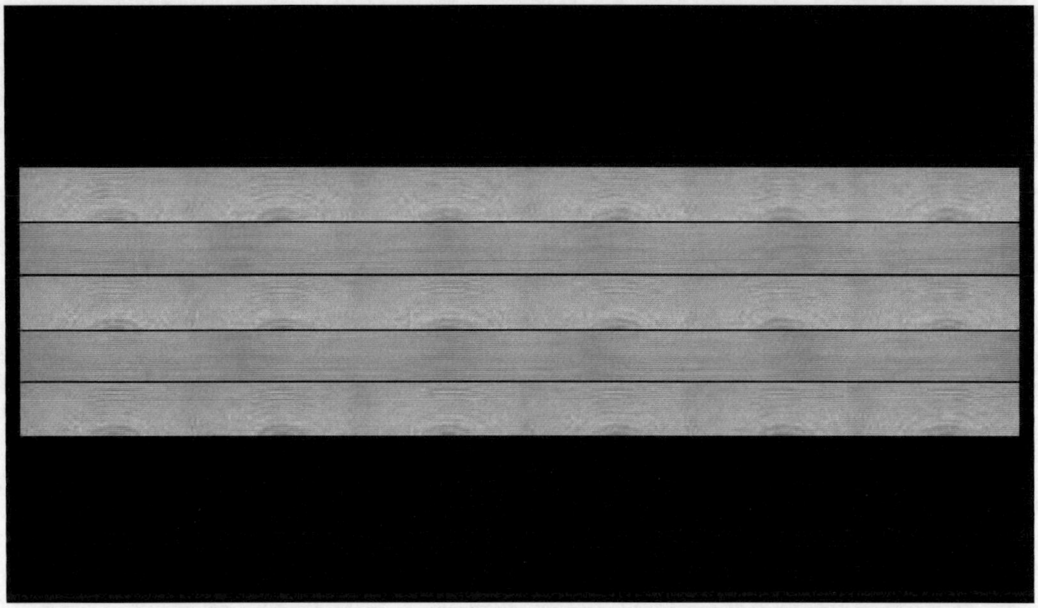

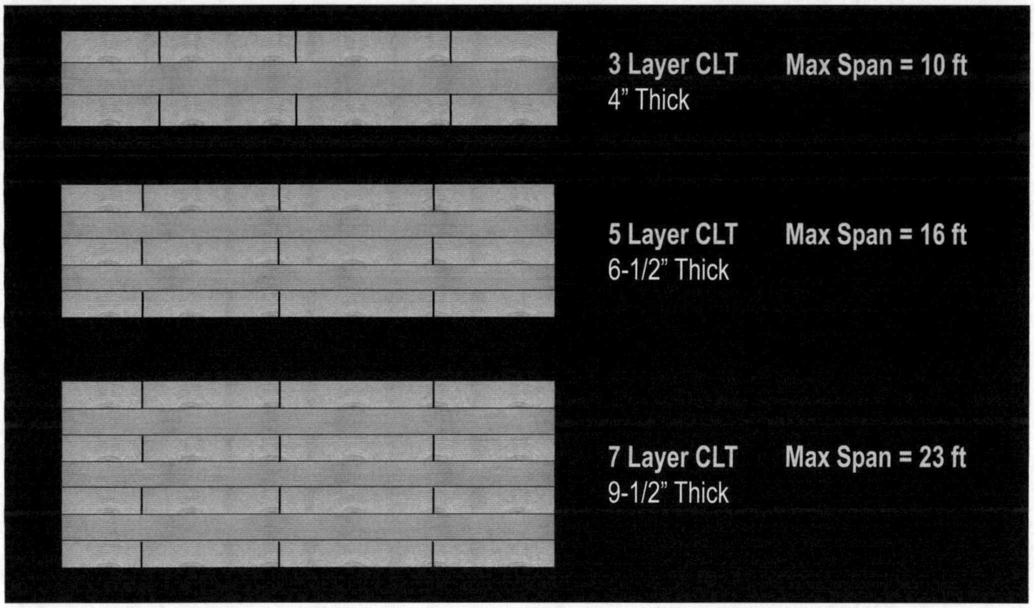
3 Layer CLT
4" Thick
Max Span = 10 ft
5 Layer CLT
6-1/2" Thick
Max Span = 16 ft
7 Layer CLT
9-1/2" Thick
Max Span = 23 ft

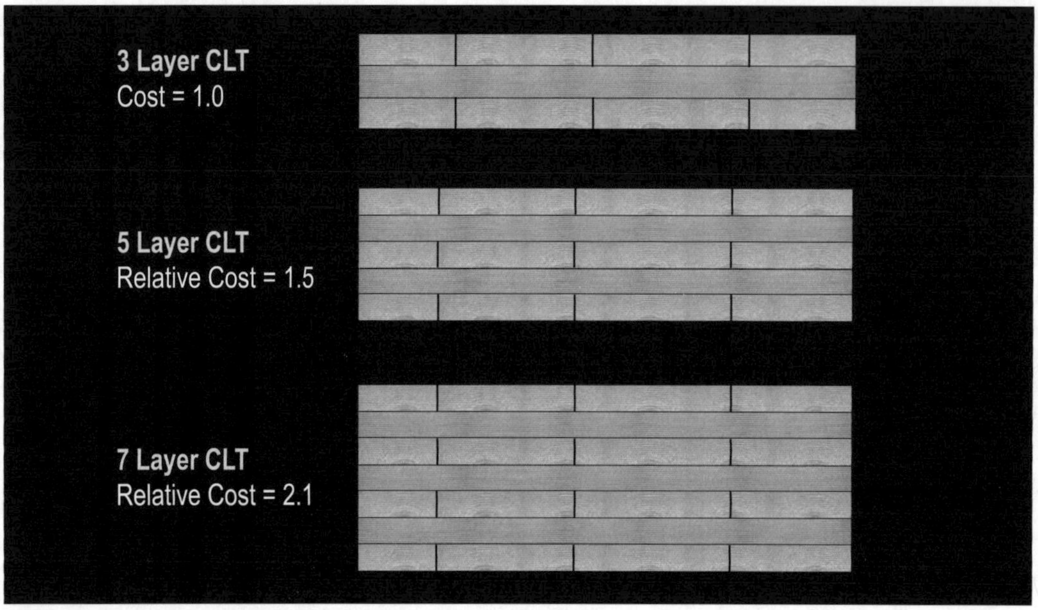
3 Layer CLT
Cost = 1.0
5 Layer CLT
Relative Cost = 1.5
7 Layer CLT
Relative Cost = 2.1

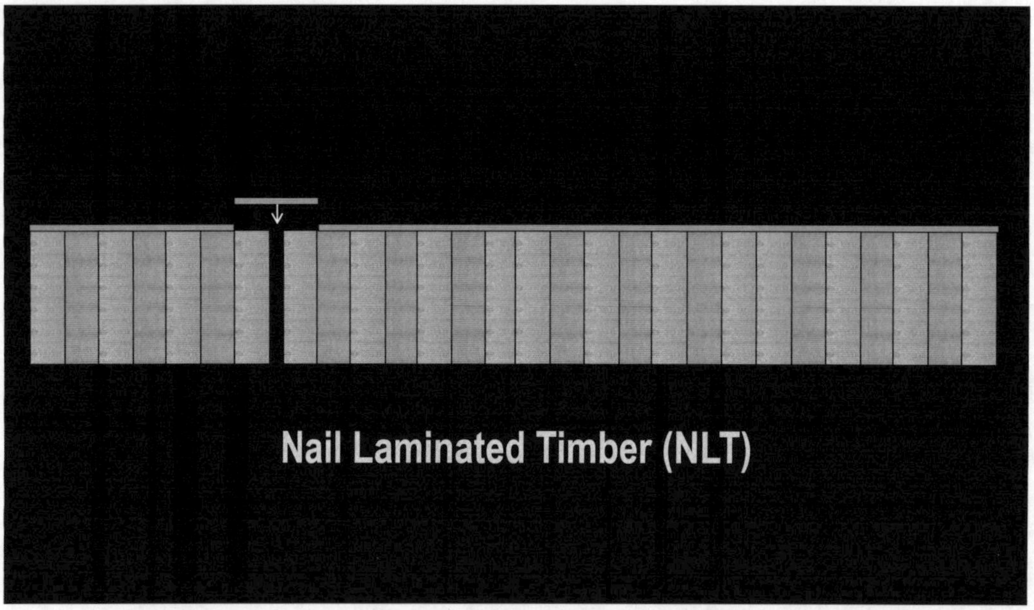
Nail Laminated Timber (NLT)

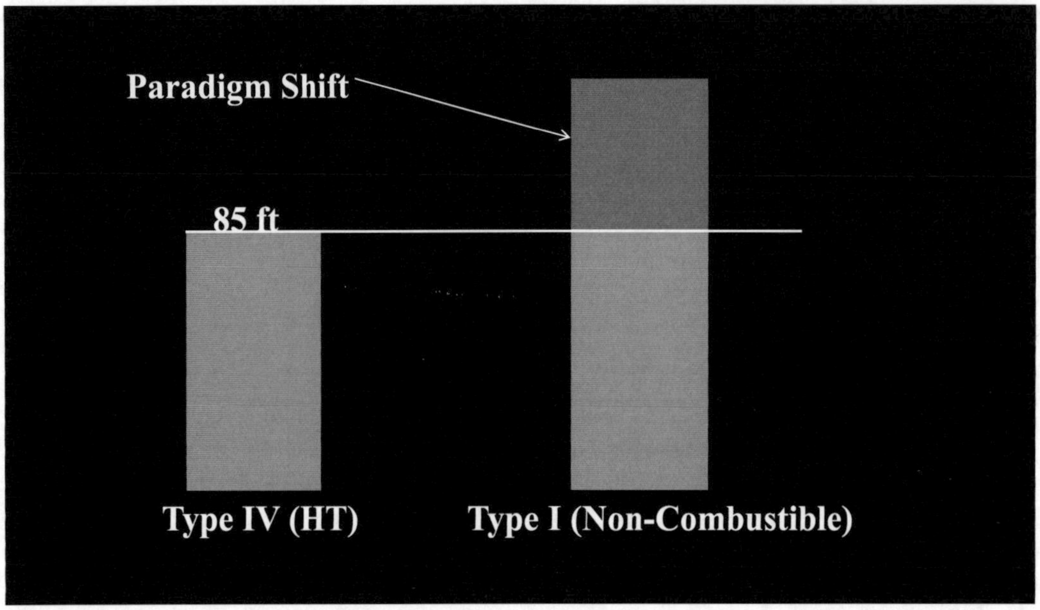
Paradigm Shift
85 ft
Type IV (HT)
Type I (Non-Combustible)

[redacted]

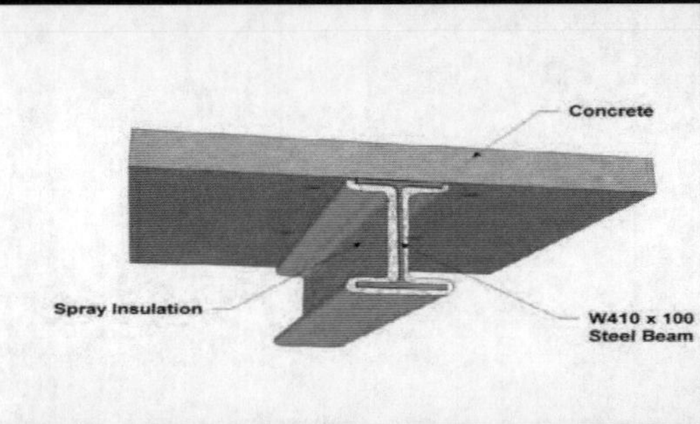

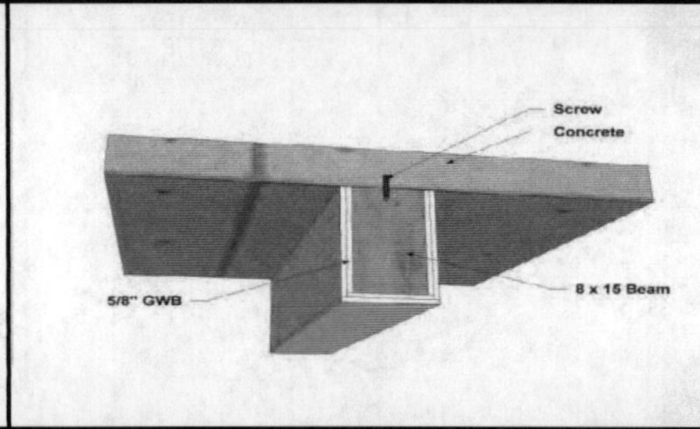

“Noncombustible construction” “Combustible construction”

[redacted]

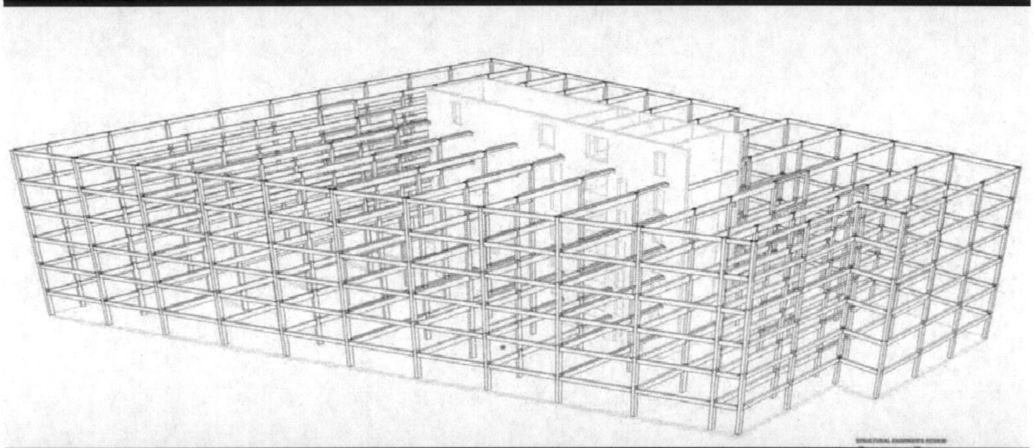

[redacted]

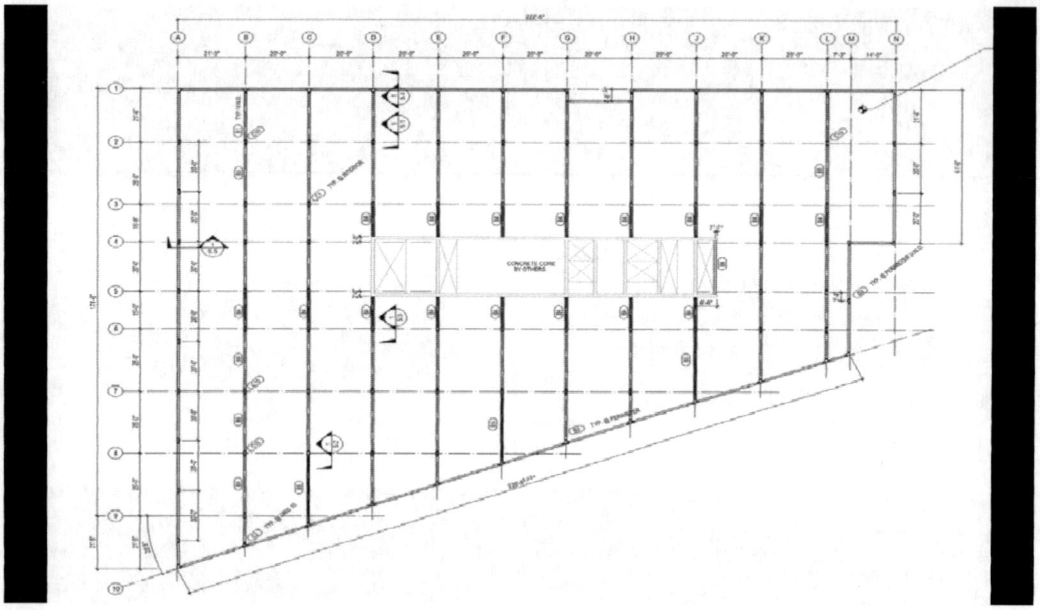

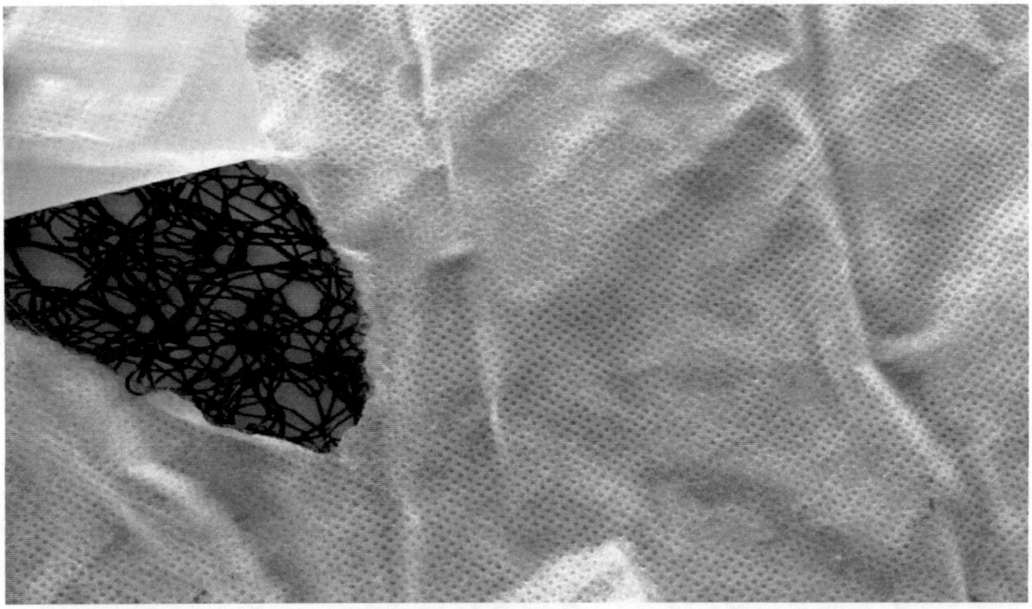

Thank You
listen.DESIGN.deliver

Ecovative
We grow materials
MycoBoard™ Introduction
Jeff Betts, Design & Development Lead

Grow Manufacturing
MycoBoard – Composite Panel with No Added Formaldhyde, no added chemicals
Green Island, New York
Founded in 2007
80 team members
Patents around Mushroom Technology in 35+ Countries

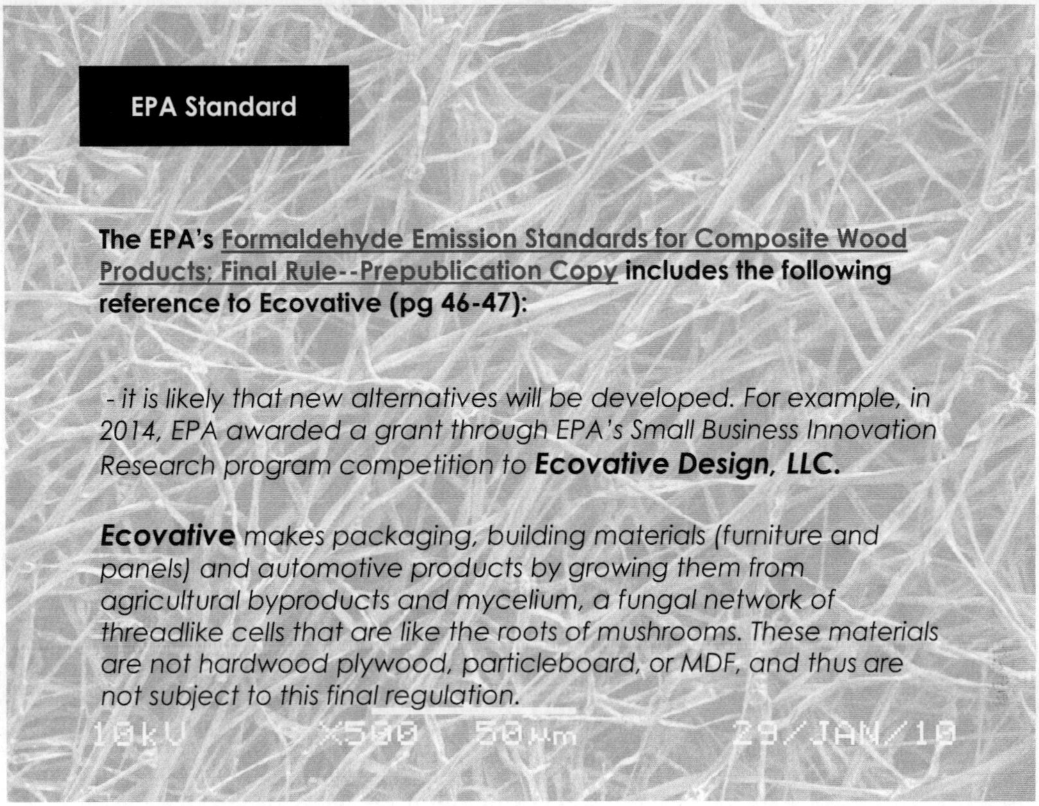
EPA Standard
The EPA's Formaldehyde Emission Standards for Composite Wood Products; Final Rule--Prepublication Copy includes the following reference to Ecovative (pg 46-47):
- it is likely that new alternatives will be developed. For example, in 2014, EPA awarded a grant through EPA's Small Business Innovation Research program competition to Ecovative Design, LLC.
Ecovative makes packaging, building materials (furniture and panels) and automotive products by growing them from agricultural byproducts and mycelium, a fungal network of threadlike cells that are like the roots of mushrooms. These materials are not hardwood plywood, particleboard, or MDF, and thus are not subject to this final regulation.
10kV X500 50µm 29/JAN/10

Mushrooms.
Mushrooms are nature's recycling system, growing on plant-based materials, self replicating without any outside energy inputs.

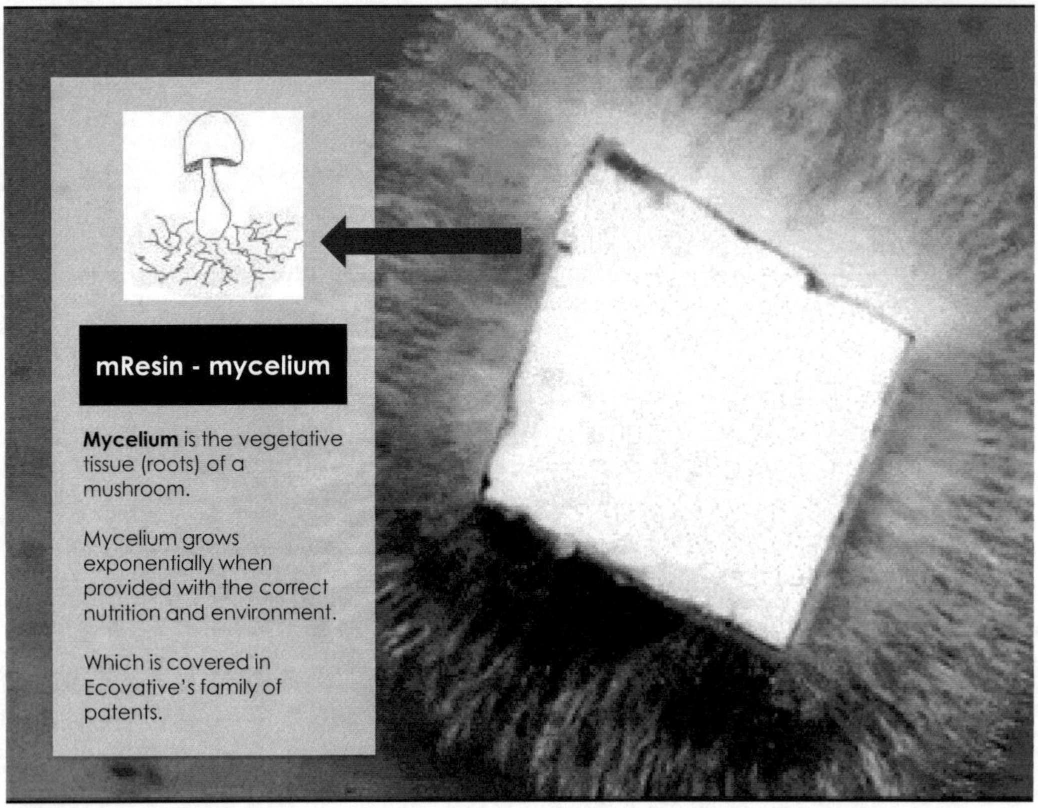
mResin - mycelium
Mycelium is the vegetative tissue (roots) of a mushroom.
Mycelium grows exponentially when provided with the correct nutrition and environment.
Which is covered in Ecovative's family of patents.

in situ resin production
Starts with ≤1% [m:m] resin
Grows to >25% [m:m] in 5 days
NONE SEI 5.0kV X100 100μm WD 11.4mm

Roots in Packaging
Protective Packaging with Sealed Air Corp
Blocking & Bracing Products like Servers for Dell, or Furniture for companies like Steelcase and Gunlocke

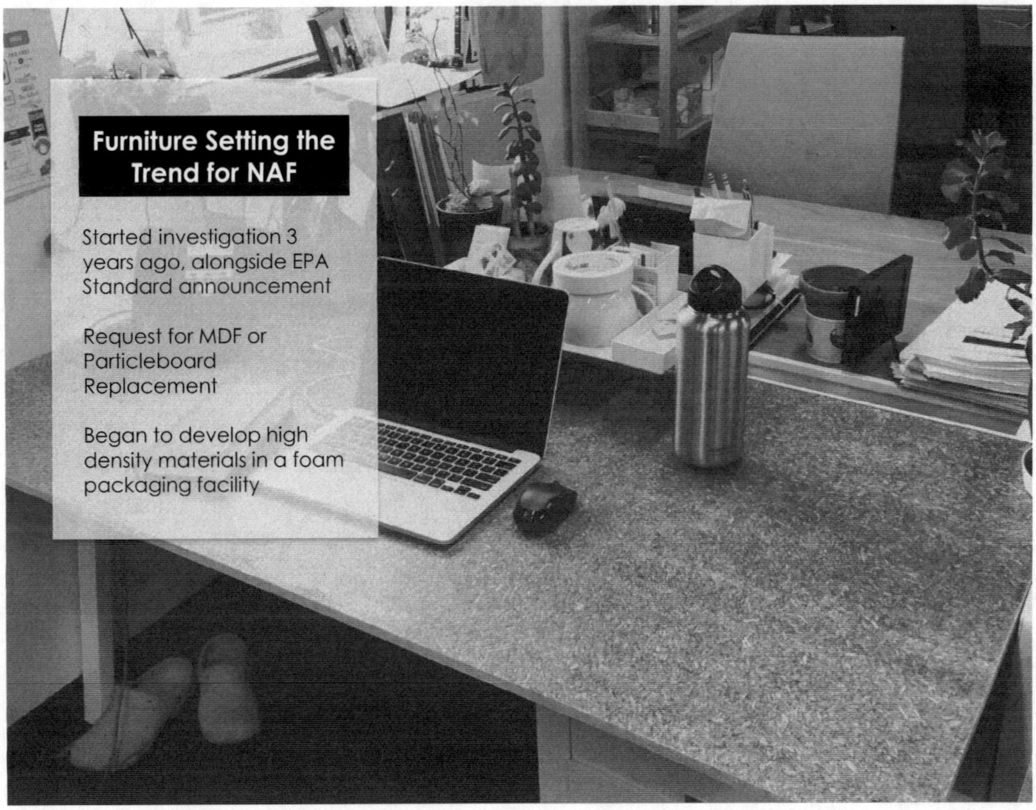
Furniture Setting the Trend for NAF
Started investigation 3 years ago, alongside EPA Standard announcement
Request for MDF or Particleboard Replacement
Began to develop high density materials in a foam packaging facility

ecovative
We grow materials.
MycoBoard
Density: 45 lb/ft^3
Screw Hold: 122 lbF
MoE: 475,000 psi
MoR: 2201 psi
IB: 84 psi

MycoBoard Components
Seating

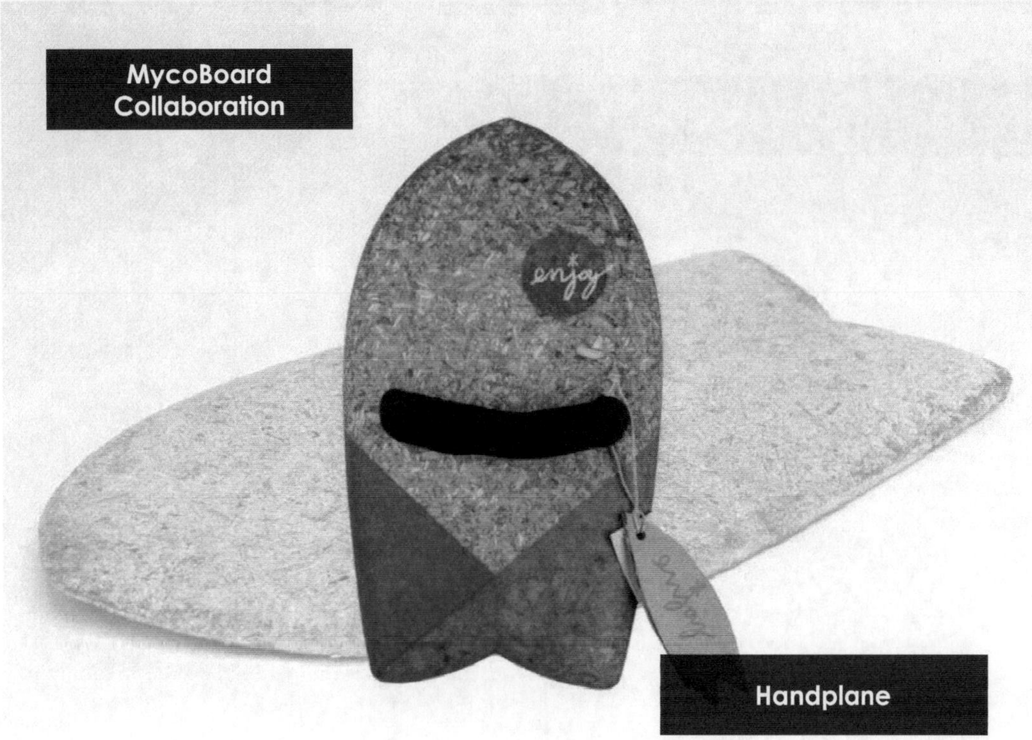
MycoBoard
Collaboration
enjoy
Handplane

Grow
Manufacturing…
… What are we
growing?

Why Mycelium ...

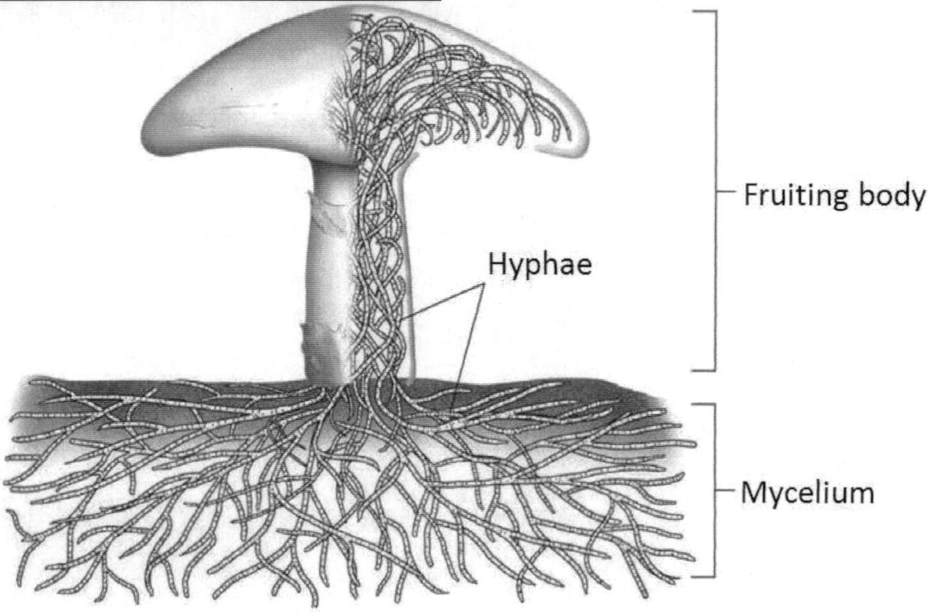

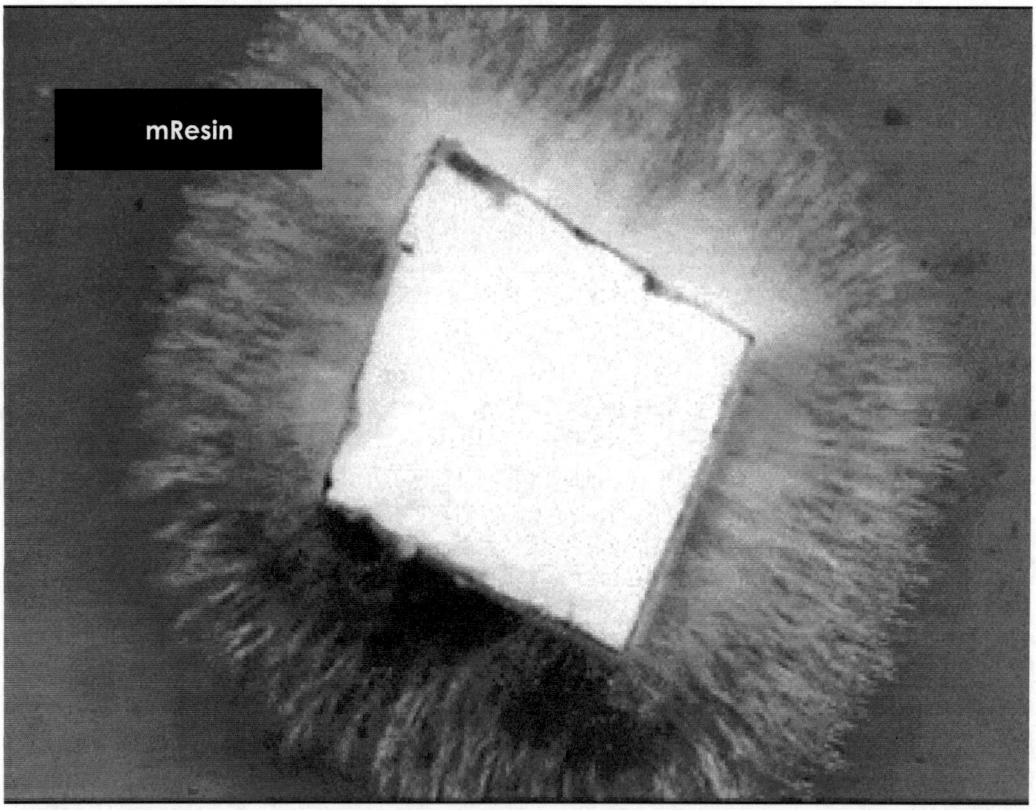

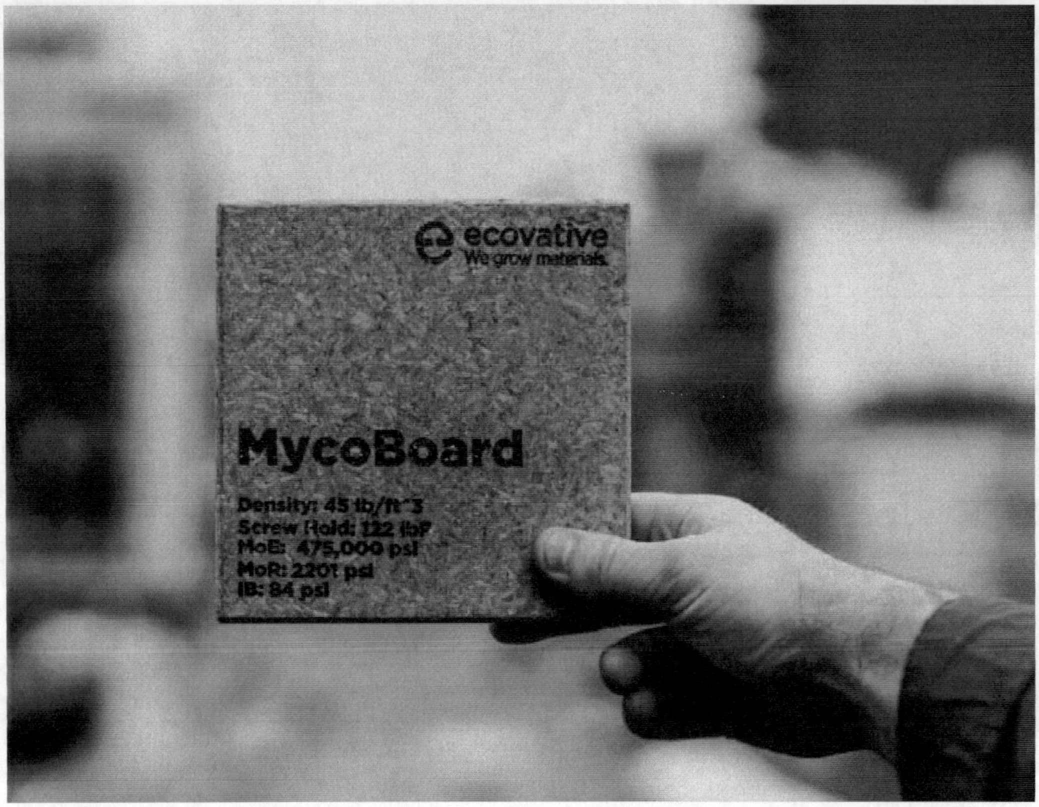
ecovative
We grow materials.
MycoBoard
Density: 45 lb/ft^3
Screw Hold: 122 lbF
MoE: 475,000 psi
MoR: 2201 psi
IB: 84 psi

Library of Inputs
Substrate
- Ag & Wood
- Sourcing
- Performance
Strain (Fungi)
- Different Performance
- More/Less Grow Time
- Greater Water Swell
- Structural Performance
- Foam mushroom is very robust
- wood mushroom even more robust
Nutrition
- Speed up growth
- Bring out certain performance

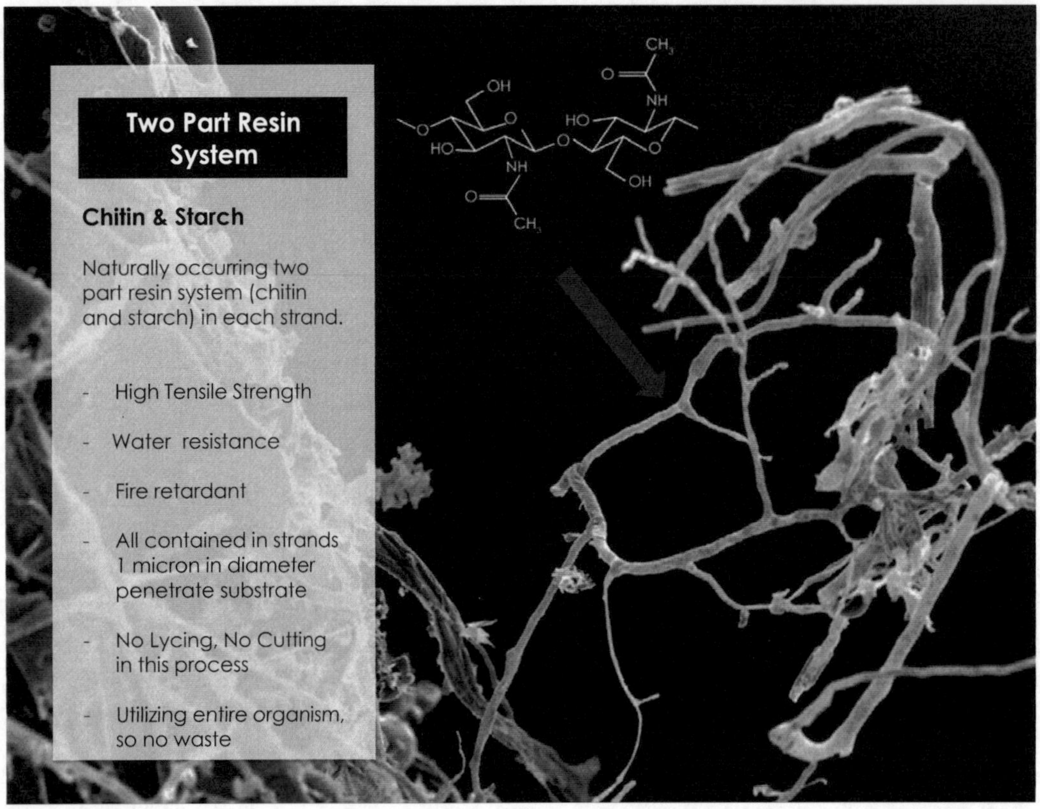
Two Part Resin System
Chitin & Starch
Naturally occurring two part resin system (chitin and starch) in each strand.
- High Tensile Strength
- Water resistance
- Fire retardant
- All contained in strands 1 micron in diameter penetrate substrate
- No Lycing, No Cutting in this process
- Utilizing entire organism, so no waste

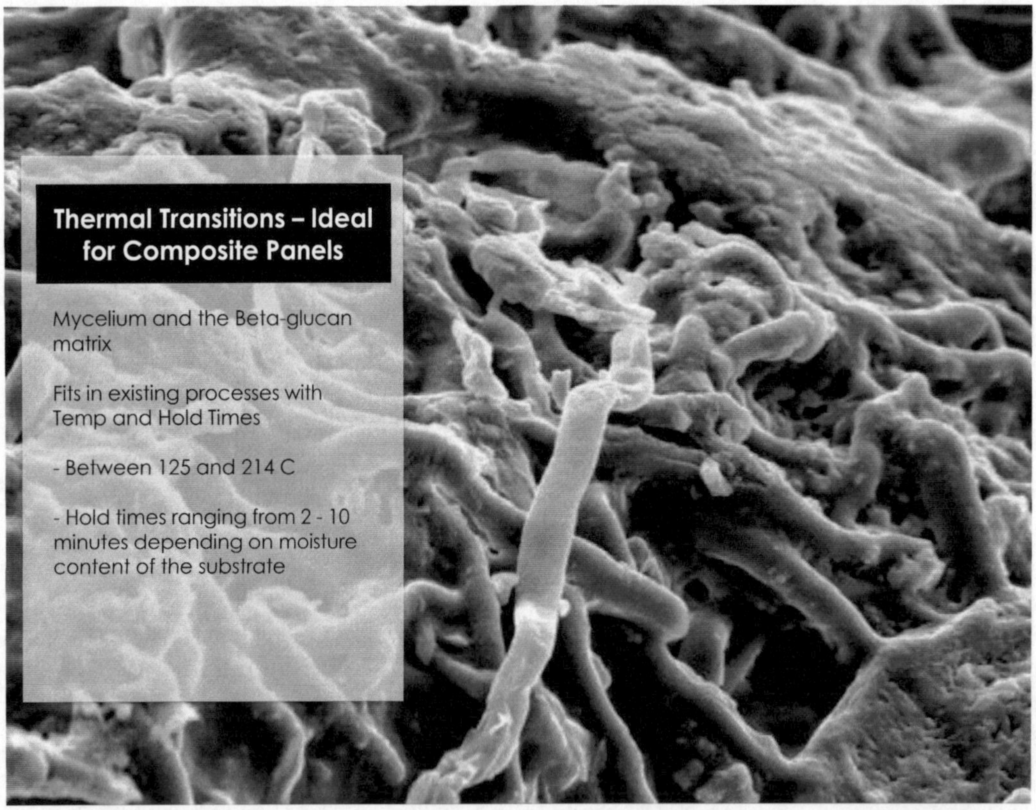
Thermal Transitions – Ideal for Composite Panels
Mycelium and the Beta-glucan matrix
Fits in existing processes with Temp and Hold Times
- Between 125 and 214 C
- Hold times ranging from 2 - 10 minutes depending on moisture content of the substrate

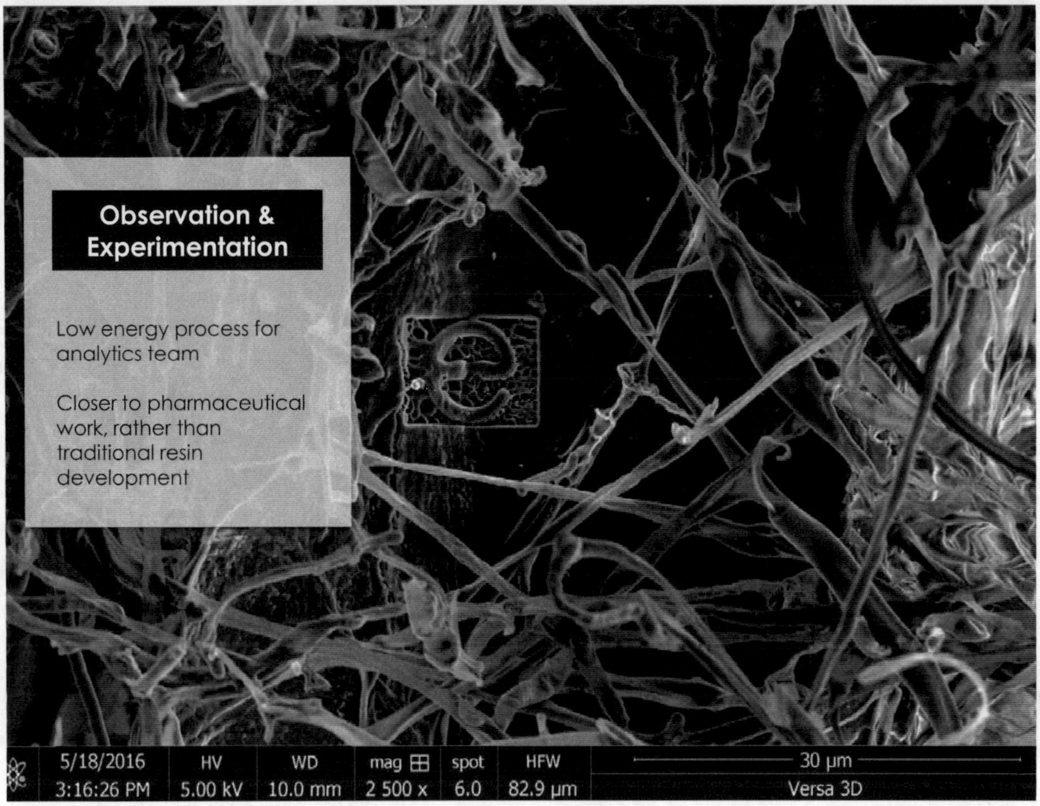
Observation & Experimentation
Low energy process for analytics team
Closer to pharmaceutical work, rather than traditional resin development
5/18/2016
3:16:26 PM
HV
5.00 kV
WD
10.0 mm
mag
2 500 x
spot
6.0
HFW
82.9 μm
30 μm
Versa 3D

Manufacturing Today
Packaging and Foam materials at 5-10 lb/ft^3
Vertical Grow System
5 Days of low energy growth
MycoBoard at 40-50 lb/ft^3

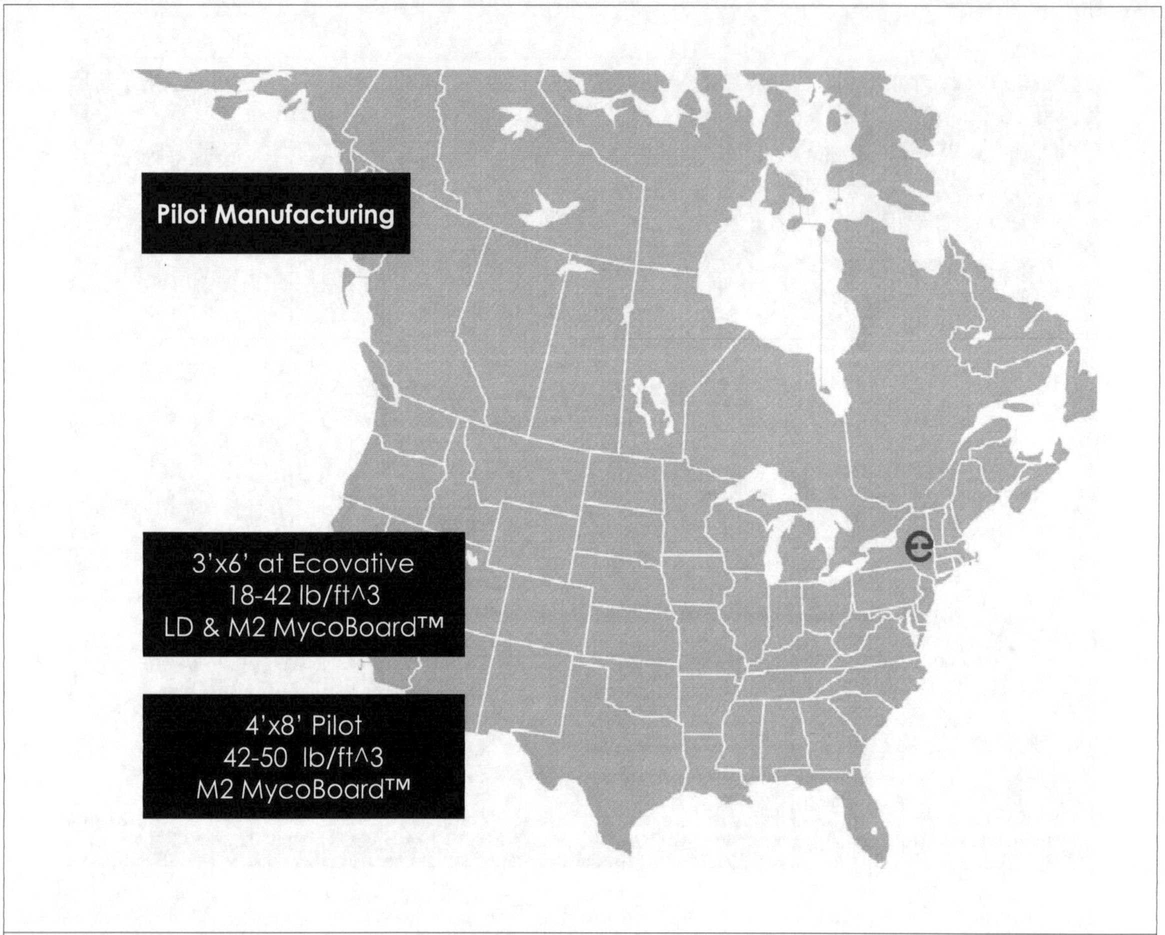
Pilot Manufacturing
3'x6' at Ecovative
18-42 lb/ft^3
LD & M2 MycoBoard™
4'x8' Pilot
42-50 lb/ft^3
M2 MycoBoard™

Scaling from Current System
40 metric tons of material in first pilot
Targeting Particleboard Performance
Ag Fiber Blend – proving what we know today

Performance

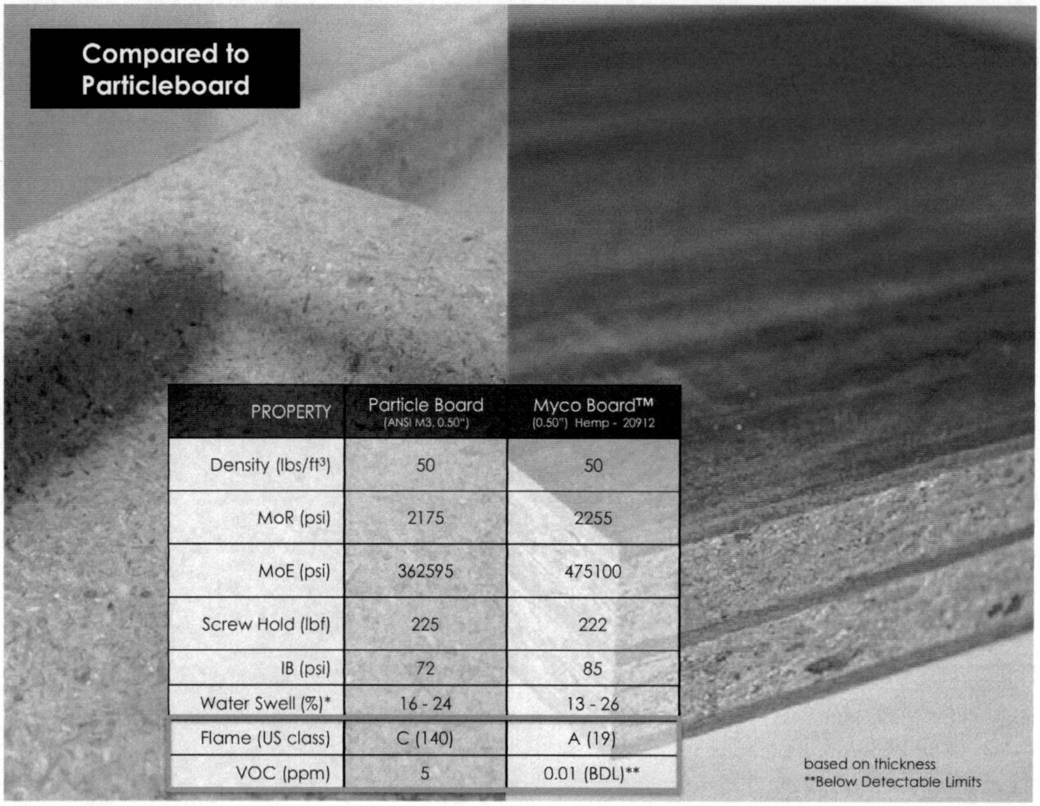

Compared to Particleboard

PROPERTY	Particle Board (ANSI M3, 0.50")	Myco Board™ (0.50") Hemp - 20912
Density (lbs/ft³)	50	50
MoR (psi)	2175	2255
MoE (psi)	362595	475100
Screw Hold (lbf)	225	222
IB (psi)	72	85
Water Swell (%)*	16 - 24	13 - 26
Flame (US class)	C (140)	A (19)
VOC (ppm)	5	0.01 (BDL)**

based on thickness
**Below Detectable Limits

Existing Feed Stocks

EU MDF Fiber   **US** Softwood Particles   **US** Soft/Hardwood Mix

PROPERTY	MDF (0.50")	Myco (0.50") MDF Fiber	M2 (0.50") PB	Myco (0.50") PB Furnish
Density (lb/ft3)	50	50	45	45
MoR (psi)	1682	1449	1885	1810
MoE (psi)	261040	300055	290100	516500
Screw Hold (lbf)	180	225	202	NT
IB (psi)	80	NT	58	75

Finishing &
Laminating

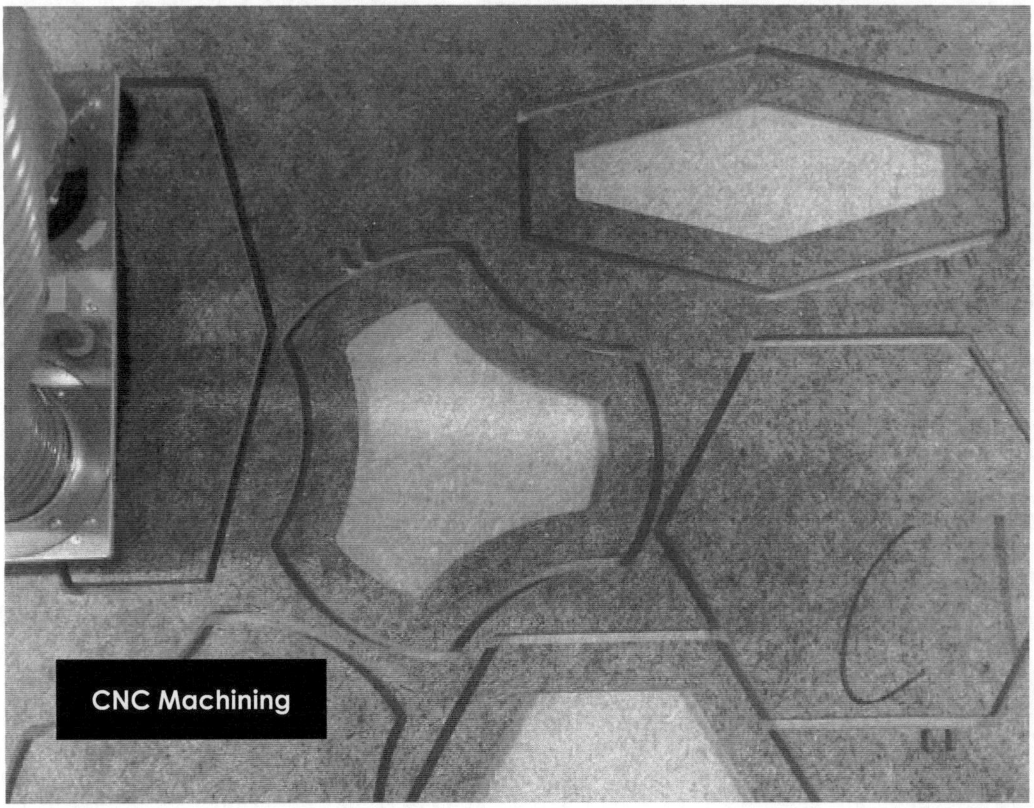
CNC Machining

Product
Manufacturing

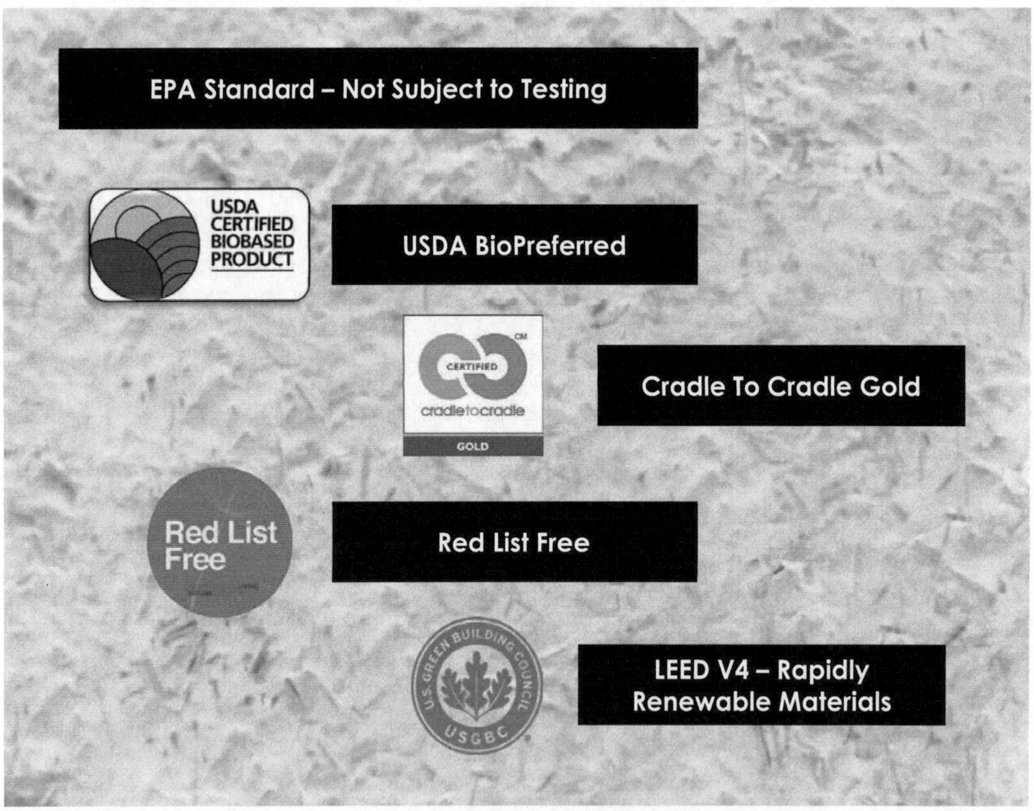
EPA Standard – Not Subject to Testing
USDA CERTIFIED BIOBASED PRODUCT
USDA BioPreferred
CERTIFIED cradletocradle GOLD
Cradle To Cradle Gold
Red List Free
Red List Free
U.S. GREEN BUILDING COUNCIL USGBC
LEED V4 – Rapidly Renewable Materials

Thank You
ecovativedesign.com
518.273.3753
Jeff Betts:
Jeffbetts@ecovativedesign.com
PS1 MoMA
2013 YAP

D.R. Johnson Wood Innovations

CLT Story to Date: Lessons Learned & the Challenges Ahead

STORY TO DATE:

Manufacturing APA/ANSI Certified CLT Panels

STORY TO DATE: **Dedicated CLT Manufacturing/Fabrication Building**

STORY TO DATE: **Utilizing Proven Manufacturing Processes**

DR Johnson
Wood Innovations
Riddle OR

STORY TO DATE:

Fabricating Panels/Beams to Tight Tolerances

STORY TO DATE:

CLT Panels To Fit Precisely Upon Arrival

STORY TO DATE:

Transporting APA/ANSI Certified CLT Panels

STORY TO DATE:

Providing On-Site Installation Assistance

STORY TO DATE:

DR Johnson Wood Innovations

STORY TO DATE:

Culminating in our Biggest Project To Date

ALBINA YARD PROJECT **4-Story Mixed-Use Portland, OR**

STORY TO DATE:

ALBINA YARD- CLT Panels (Floor) & Glulam Beams/Columns

STORY TO DATE:

Project Pipeline- GREAT THINGS TO COME

LESSONS LEARNED: **Not Just a Materials Supplier**

LESSONS LEARNED: **PROJECT TEAM COLLABORATION**

Architects/ Engineers/ Developers- Pre Construction

DRJohnson
Wood Innovations
Riddle OR

Attend Project Design Meetings

Review Plan Sets at Various Stages of Design

Provide Budgetary Cost Estimates at each Stage of Design

Building Department Review/ Product Testing

LESSONS LEARNED: **PROJECT TEAM COLLABORATION**

Architects/General Contractors- Project Awarded

Contracts/ Payment Terms & Performance Bonds

3-D Modeling- Collison Checks- Panel/Beam Layouts

Product Testing

Shop Drawing Process

LESSONS LEARNED: **PROJECT TEAM COLLABORATION**

Installers- Construction Begins

Now we Make the CLT Panels

Collaborate on Installation Schedule/ Panel Order

Collaborate on Staging & Off-Loading Trucks

LESSONS LEARNED: **Technically Capable Workforce**

LESSONS LEARNED: **Making CLT marketable**

LESSONS LEARNED: **Product Testing**

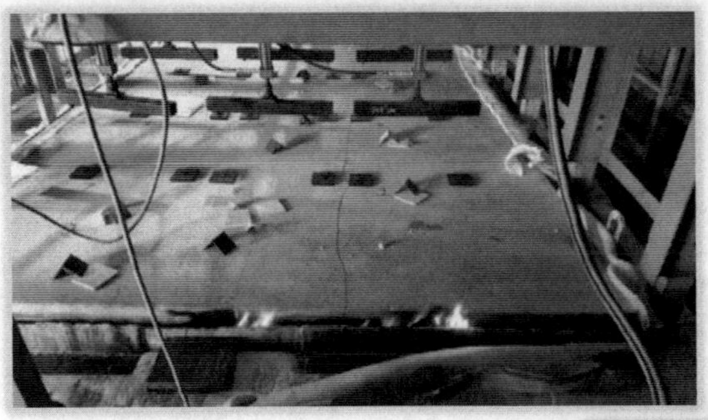

ASTM E119 TEST

2-Hour Rating Achieved

DR Johnson Wood Innovations Riddle OR

LESSONS LEARNED: **Product Testing**

DR Johnson Wood Innovations Riddle OR

ASTM E84 TEST

FLAME SPREAD RATING- A

GOING FORWARD: **Collaboration & Standardization**

Overview of Environmental Regulations Facing the Industry

Jackson Morrill
President
Composite Panel Association
Wood Composites Symposium
Atlanta, GA
August 23, 2016

COMPOSITE PANEL ASSOCIATION

1

Profile of the Composite Panel Industry

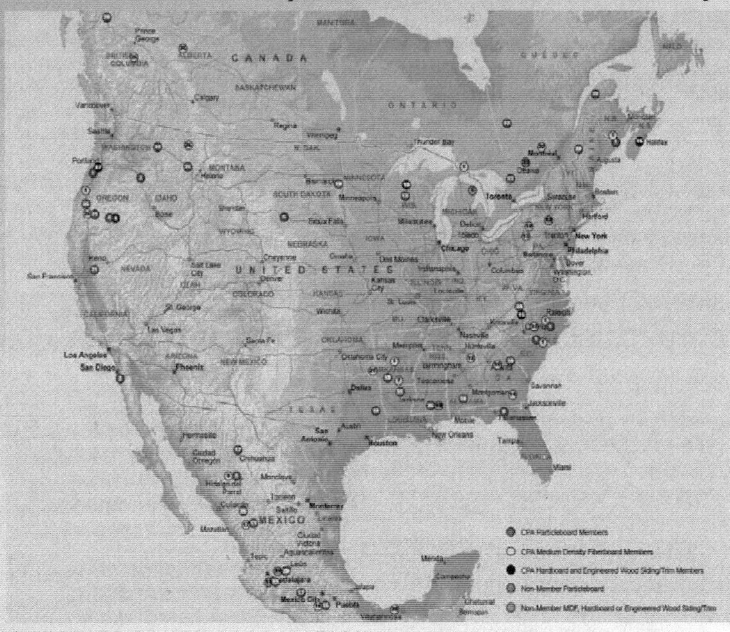

COMPOSITE PANEL ASSOCIATION

Regulatory Update

- EPA Formaldehyde Regulation
- CARB Amendments to ATCM
- Lumber Liquidators Developments
- US Clean Power Plan

COMPOSITE PANEL ASSOCIATION

3

U.S. EPA Formaldehyde Regulation

- Congress passed law in 2010 (TSCA Title VI)
  - Supported by NGO, union and industry stakeholders
  - Called for regulations by Jan. 1, 2013
  - Specified emission limits, test methods to align with CARB 2
- Prepublication version published (7/27/16)
  - Awaiting publication in Federal Register to start timeline for implementation

COMPOSITE PANEL ASSOCIATION

4

Key Provisions – CARB Alignment

- Same product coverage – HWPW, PB, MDF
- Same exemptions
  - Hardboard
  - Softwood plywood
  - OSB
  - Various structural wood products
  - Certain wood windows and doors
- Same emission limits
  - HWPW – 0.05 ppm
  - MDF – 0.11 ppm
  - Thin MDF – 0.13 ppm
  - PB – 0.09 ppm

COMPOSITE PANEL ASSOCIATION

5

Key Provisions – CARB Alignment

- Rigorous TPC system
- Reduced testing and no third-party certification required for "exempt" ULEF and NAF
- Same test methodologies
- Similar methods for recognition of equivalency and correlation
- Similar labeling requirements
- Similar reporting requirements
- Similar record-keeping requirements

COMPOSITE PANEL ASSOCIATION

6

Regulation of Laminated Products

- Defines laminated products (LPs)
  - Wood or woody grass veneers only
  - Component part used in construction or assembly of finished good
- Composite wood using other veneers, papers, etc. are not LPs
  - Considered "finished goods"
  - Must document compliant core w/in 1 year of effectiveness

COMPOSITE PANEL ASSOCIATION

7

Two Types of Laminated Products

- Type #1 – panels where veneer attached to substrate with NAF or PF resin
  - Only recordkeeping required
- Type #2 – panels that use any other resin to attach veneer (e.g., UF, ULEF)
  - Meet 0.05 ppm for HWPW
  - Comply with certification, recordkeeping and testing
- Timing
  - Compliant core – 1 year
  - All else – 7 years

COMPOSITE PANEL ASSOCIATION

8

Other Provisions

- De Minimis Exemption
  - New – not in CARB
  - Exempt from labeling - 144 square inches
  - Must use compliant composite wood products
  - Comply with record-keeping
- No designated sell-through provisions
  - Panel made prior to "manufactured-by" date is compliant
  - Maintain records of the date of manufacture

COMPOSITE PANEL ASSOCIATION

9

California Air Resources Board

- Amendments to ATCM expected in 2018
  - Question whether will align with EPA Regulation
- Deconstructive testing
  - Exploring inter-laboratory comparison work to address concerns over repeatability and reproducibility
- Additional enforcement?

CARB PHASE 2 COMPLIANT

COMPOSITE PANEL ASSOCIATION

10

Lumber Liquidators - CPSC

- **CPSC investigation (3/15-6/16)**
  - CDC Report on health effects of tested samples
  - CDC/ATSDR Website
  - CPSC announced end to investigation
  - LL agrees to:
    - "Not resume sales of laminated wood flooring previously imported from China"
    - Conduct testing as part of 3-year program for Chinese-made laminate flooring sold from Feb. 2012-May 2015

COMPOSITE PANEL ASSOCIATION

11

Lumber Liquidators – CA Regs

- CARB Enforcement (3/22/16)
  - $2.5 million settlement
  - Audit program
  - "Finished goods sample" testing for high risk products
- Prop 65 Lawsuit (4/4/16)
  - LL was sued for failure to warn in accordance with CA Prop 65
  - State court judge dismissed claims

COMPOSITE PANEL ASSOCIATION

12

Wood Fiber Issue

"How Europe's climate policies led to more U.S. trees being cut down" The Washington Post

"A Biofuel Debate: Will Cutting Trees Cut Carbon?" The New York Times

BIOMASS101

http://www.biomass101.org/

"The world's most reusable resource is getting even greener."

http://www.nrdc.org/energy/forestsnotfuel/

Clean Power Plan

- Sets emission targets for carbon pollution from power plants
  - Reduce 32% by 2030
  - Guidelines for states to develop plans to meet state specific goals set forth by EPA

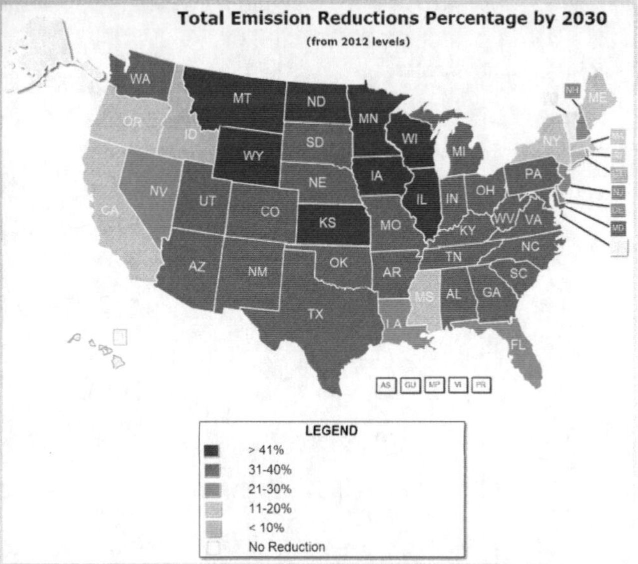

Source: NCSL (2015)

COMPOSITE PANEL ASSOCIATION

14

Biomass Under CPP

- Not all biomass is carbon neutral
  - EPA retains authority to determine what biomass counts towards reduction targets ("qualified biomass")
  - Alternative fate of feedstock is considered (i.e., leakage)
- CPA has argued that we are an "existing use," therefore our industry residuals should not be qualified

Chips

Sawdust

Shavings

Trim

COMPOSITE PANEL ASSOCIATION

15

Clean Power Plan Timeline

15 Years	• January 1, 2030 - CO_2 Emission Goals met
Summer 2015	• August 3, 2015 - Final Clean Power Plan
1 Year	• September 6, 2016 - States submit initial state plan
3 Years	• September 6, 2018 - States submit final state plan
7 Years	• January 1, 2022 - Compliance period begins

COMPOSITE PANEL ASSOCIATION

16

EPA Clean Power Plan

Timing – delayed!

- Supreme Court issued decision to stay the CPP (2/9/16)
- 30 states filed case against federal government in DC Circuit Court of Appeals on the merits
  - Court to hear case in September
- Case may be referred to Supreme Court (Scalia vacancy)
- Should CPP eventually go forward, deadlines for SIPs would likely be delayed by amount equal to SCt.'s stay

COMPOSITE PANEL ASSOCIATION

17

Thank you!

COMPOSITE PANEL ASSOCIATION

Mechanical, Acoustic, and Fire Properties of Southern Pine Cross-Laminated Timber

Hindman, D*., Bouldin, J*., Loferski, J., Bond, B., and Quesada-Pineda, H.J.

IWF 2016

*Lead project investigators

SUSTAINABLE BIOMATERIALS

VirginiaTech Invent the Future

Agenda

- Uses of cross-laminated timber (CLT)
- Research on southern yellow pine (SYP) and CLT
  - Mechanical, acoustic, and fire properties
- Conclusion

SUSTAINABLE BIOMATERIALS

VirginiaTech Invent the Future

Applications of CLT

- What is CLT?
  - Cross: each layer is "crossed" or runs 90 degrees to previous layer
  - Laminated: Each layer is typically 2x (1.5 inches rough, 1.375 dressed) and bonded with adhesive
  - Timber: Creates a large section of wood usually used as panel

SUSTAINABLE BIOMATERIALS

VirginiaTech *Invent the Future*

Applications of CLT

- What is massive timber?
  - Building System Using Large Panelized Wood Elements
  - Similar to Pre-Cast Concrete Construction
  - Mass Timber Uses
    - Laminated Veneer Lumber
    - Glue Laminated Lumber
    - Nail Laminated Lumber
    - Cross Laminated Lumber: **New Addition**

Franklin Elementary School, WV. By Charles Judd.

SUSTAINABLE BIOMATERIALS

VirginiaTech *Invent the Future*

Applications of CLT

- What is the Innovation?
  - Common Materials – Wood and Glue
  - Building Methods are the True Innovation
  - Eliminates need for Joists / Interior Walls
  - Panels are 'Self-Supporting'
  - Large Racking Strength of Panels
  - Structural Form Is Simplified

Franklin Elementary School, WV. By Charles Judd.

SUSTAINABLE BIOMATERIALS

VirginiaTech
Invent the Future

Applications of CLT

- Recent CLT structures
  - Integrated Design Building, UM at Amherst, MA
    - Designed by Leers Weinzapfel Associates

SUSTAINABLE BIOMATERIALS

VirginiaTech
Invent the Future

Applications of CLT

- Recent CLT structures
  - Franklin School Project, West Virginia
    - Built by Charles Judd, Blue Heron Construction, Inc.

SUSTAINABLE BIOMATERIALS

VirginiaTech
Invent the Future

Applications of CLT

- WoodTek Headquarter, Taiwan
  - By Equilibrium

SUSTAINABLE BIOMATERIALS

VirginiaTech
Invent the Future

Virginia Tech's Connection

- Collaboration with Southern Virginia Higher Education Center (SVHEC) to investigate use of SYP for CLT
- USDA grant to investigate potential of Low-value hardwood lumber for CLT
- Our Approach has been to Collaborate to provide the Wood Material and Engineering Experience to Support our Partners

SUSTAINABLE BIOMATERIALS

VirginiaTech Invent the Future

CLT and SYP

- Research objectives
- Demonstrate that SYP CLT panels meet performance standards for use in commercial construction in the United States
  - Structural Capacity – Comparison to PRG 320 values
  - Fire Capacity – Demonstration of resistance to fire needed
  - Acoustic Capacity – Essential for use in multi-family residences

SUSTAINABLE BIOMATERIALS

VirginiaTech Invent the Future

CLT manufacturing

- 5 layer CLT panels
- All layers No. 2 Southern Pine
- PU adhesive from several companies
- Clamping fixture to edge glue each layer
- CNC flycutter used for surfacing
- Cold press for panel consolidation

SUSTAINABLE BIOMATERIALS

VirginiaTech Invent the Future

Bending Specimen Manufacture

- Maximum panel length was 1.98 m
- Developed a multiple lap joint 15.2 cm long for connection of three panels
- CNC used for cutting lap joint
- PU adhesive used for a re-pressing of joint
- Additional screw reinforcement added for testing

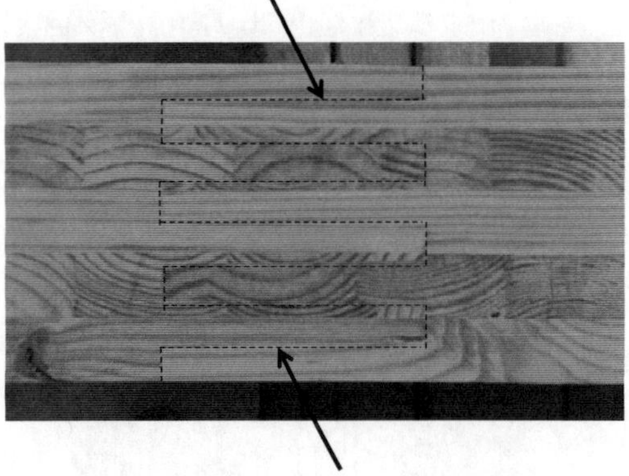

SUSTAINABLE BIOMATERIALS

VirginiaTech Invent the Future

Bending Testing

- Based on ASTM D 198 and PRG-320 recommendations (h = 175 mm)
- Bending Strength / Stiffness Evaluation
  - L/h = 30
  - L = 5.23 m, Actual span = 4.72 m

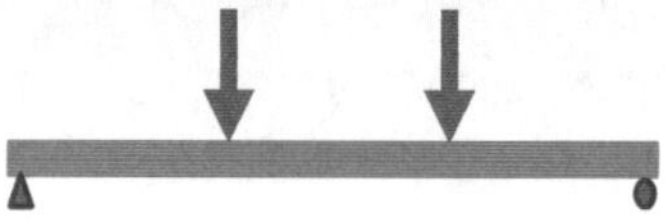

SUSTAINABLE BIOMATERIALS

VirginiaTech Invent the Future

Shear Testing

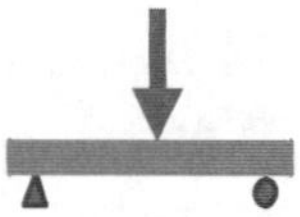

- Shear Strength
  - L/h = 5 to 6
  - L = 0.873 to 1.05 m, Actual span = 1.07 m
- Reached maximum load on test machine (~200 kN)
- Reduced width to 22.9 cm

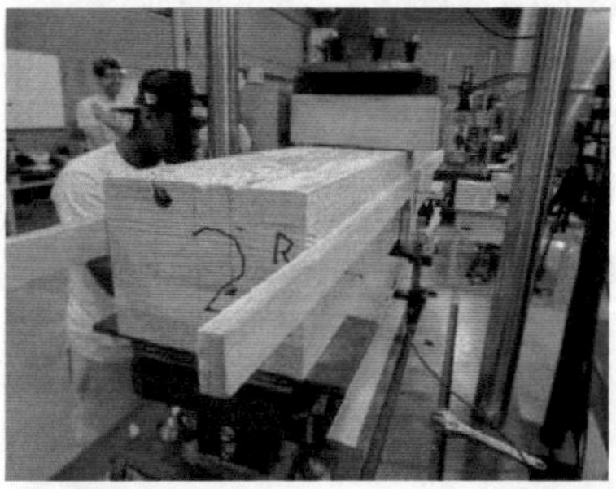

SUSTAINABLE BIOMATERIALS

VirginiaTech Invent the Future

Subsequent Tests

- Bond Line Shear Strength – AITC T107
- Moisture Content – ASTM D 4442 (Oven Dry Method)
- Specific Gravity – ASTM D 2395 (Volume by Immersion)
- Cyclic Delamination Test – AITC T110

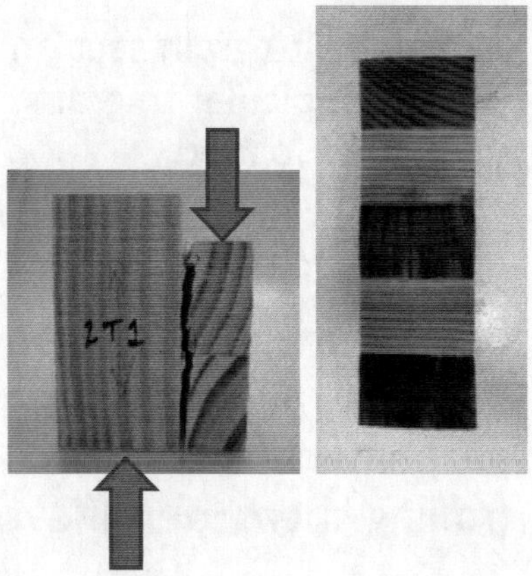

SUSTAINABLE BIOMATERIALS

VirginiaTech *Invent the Future*

Mechanical Property Results

Property	Experimental Value	PRG-320 / NDS Value
Allowable Bending Strength, F_bS	37.1 kN/m (14.3% COV)	23.1 kN/m
Bending Stiffness, EI_{eff}	4,110 kN-m^2/m (8.8% COV)	3,900 kN-m^2/m
Allowable Shear Strength, F_vA	146 kN/m (8.3% COV)	N/A
Bondline Shear Strength	4.38 MPa (3.3% COV)	N/A
Bond Line % Wood Failure	81.6% (8.2% COV)	> 80%
Cyclic Delamination Bondline Failure	17.2% (72.9% COV)	< 5%
Moisture Content	9.8% (4.4% COV)	> 8.0%
Specific Gravity	0.55 (3.1% COV)	0.55 (*NDS*)

SUSTAINABLE BIOMATERIALS

VirginiaTech *Invent the Future*

Recommendations for Future CLT Testing – Bond Line Tests

- Specification of grain orientation
  - Parallel to grain section fixed, perpendicular to grain section loaded
  - Consistent with work by Kim et al. (2013)
- Characterization of wood failure as parallel or perpendicular to grain (rolling shear identification)

SUSTAINABLE BIOMATERIALS

VirginiaTech
Invent the Future

Recommendations for Future CLT Testing – Face Delamination

- Sample is different than glulam since each face contains a parallel – perpendicular interface
- Measurement of bond line length before testing different than bond line length after testing due to specimen barreling
  - Suggest using bond line length AFTER testing

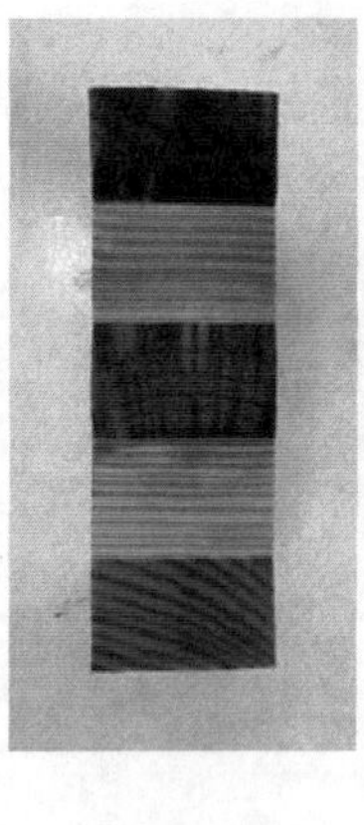

SUSTAINABLE BIOMATERIALS

VirginiaTech
Invent the Future

Acoustic Testing

- Conducted at ATI in York, PA
- ASTM E 90 Standard Test Method for Laboratory Measurement of Airborne Sound Transmission Loss of Building Partitions and Elements
- ASTM E 492 Standard Test Method for Laboratory Measurement of Impact Sound Transmission Through Floor-Ceiling Assemblies Using the Tapping Machine
- Wall Tests
  - Bare CLT
  - One side Metal Hat Channel, Rubber Isolators, R-19 batt insulation, GWB, Duct Tape
- Floor Tests
  - Bare CLT
  - One side 2x8 Joists, Rubber Isolators, R-30 batt insulation, GWB, Duct Tape

SUSTAINABLE BIOMATERIALS

VirginiaTech Invent the Future

Acoustic Results

- Wall
  - Bare CLT STC = 43
  - One-sided CLT STC = 54
- Floor
  - Bare CLT STC = 43, IIC = 27
  - One-sided STC = 56, IIC = 46
- IBC Standard: STC = 50, IIC = 50

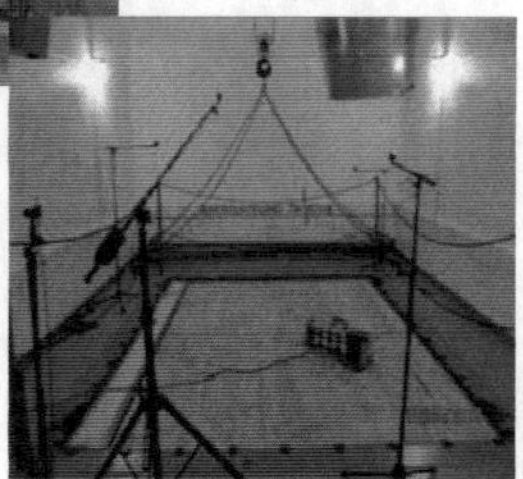

SUSTAINABLE BIOMATERIALS

VirginiaTech Invent the Future

Wall Results

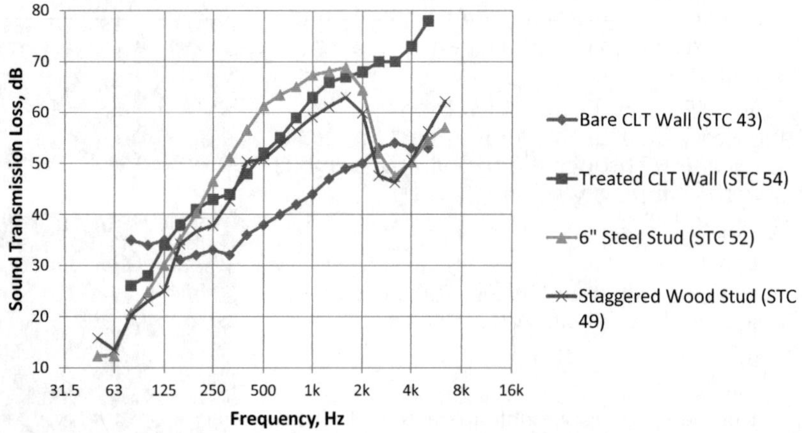

SUSTAINABLE BIOMATERIALS

VirginiaTech Invent the Future

Floor Results - STC

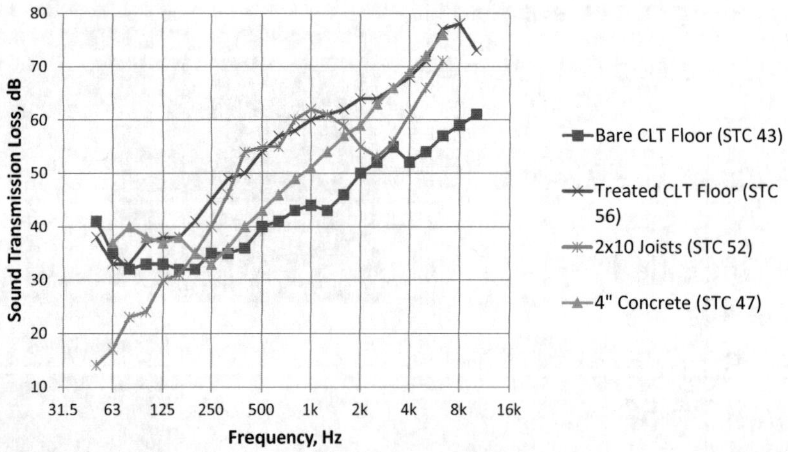

SUSTAINABLE BIOMATERIALS

VirginiaTech Invent the Future

Floor Results - IIC

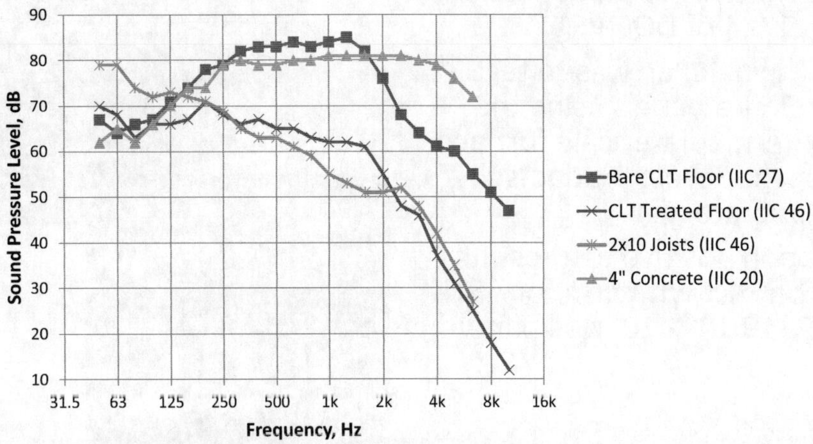

SUSTAINABLE BIOMATERIALS

VirginiaTech Invent the Future

Acoustic Testing Conclusions

- STC values from both the wall and floor with one sided treatment exceeded the IBC requirement
- Similarities in bare CLT and concrete floors (mass law)
- Similarities in treated CLT and wood walls / floors
- IIC performance similar to wood, better high frequency performance than concrete

SUSTAINABLE BIOMATERIALS

VirginiaTech Invent the Future

Fire Testing

- PRG 320 requires CLT adhesives to meet Section 6.1.3.4 of DOC PS1
- Original plan was to test 3' x 3' fire panels using the intermediate scale furnace at the Forest Products Laboratory
- Upon positive test results, contract with firm for ASTM E 119 10' x 10' wall panel

SUSTAINABLE BIOMATERIALS

VirginiaTech Invent the Future

Fire Results

- Test 1 – Bare CLT
- Test 2 – 5/8" GWB
- Test 3 – 5/8" GWB w/ spacers
- Test 4 – 2-5/8" GWB w/ spacers
- Test 5 – Intumescent
- Test 6 – 3-5/8" GWB

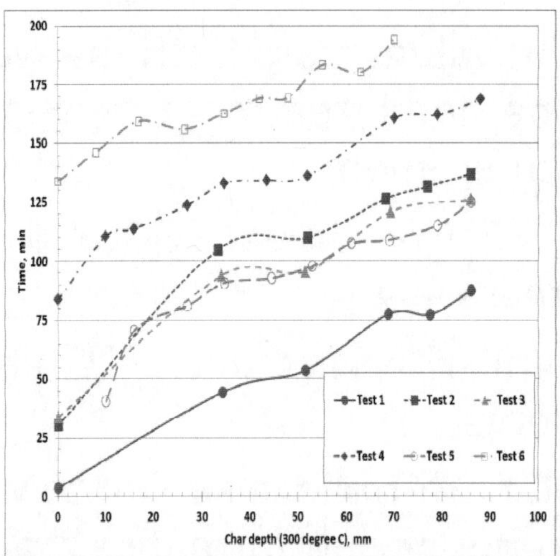

Figure 10. Times for 300°C at different depths within the CLT panels.

SUSTAINABLE BIOMATERIALS

VirginiaTech Invent the Future

Intermediate Fire Testing

Test No.	Estimated % Design Load for		
	1 Hr. Rating	2 Hr. Rating	3 Hr. Rating
1 – Bare CLT	14%	-	-
2 – 5/8" GWB	52%	14%	-
3 – 5/8" GWB w/spacer	48%	14%	-
4 – 2-5/8" GWB	100%	27%	-
5 – Intumescent	44%	11%	-
6 – 3-5/8" GWB	100%	100%	14%

SUSTAINABLE BIOMATERIALS

VirginiaTech *Invent the Future*

Adhesive Performance In Fire

- Notice bare wood under char layer which has fallen
- Researchers heard layers falling off in furnace before burn through
- Little data available on PU performance with Southern Pine at high temperatures

SUSTAINABLE BIOMATERIALS

VirginiaTech *Invent the Future*

General Conclusions

- Mechanical Performance
  - Good bending strength and stiffness compared to Grade V3
  - Did not pass cyclic delamination test
- Acoustic Performance
  - CLT wall and floor with one-sided treatment exceeded STC values in IBC
  - Floor IIC values did not meet IBC requirements
- Fire Properties
  - Assume panels can meet a two-hour fire rating with 2 layers of 5/8" GWB
  - Concerns over adhesive performance under high heat

SUSTAINABLE BIOMATERIALS

VirginiaTech Invent the Future

Current Work

- Bending and Shear Stiffness Terms of CLTs
- Dowel Embedment Strength of Multiple Layers
- Use of Low-Grade Yellow Poplar As Feedstock for CLTs
- Development of CLT Structural Modeling Capacity
- CLT Connection Design Using the NDS
- Influence of Grain Direction on Bondline Shear Tests

SUSTAINABLE BIOMATERIALS

VirginiaTech Invent the Future

Acknowledgements

- CIT Commonwealth Research Commercialization Fund
- David DeVallance, West Virginia University
- Firefree
- Henkel
- Southern Virginia Higher Education Center – David Kenealy, Travis Buchanan, Kevin Christy
- Virginia Tech – Joe Loferski, Earl Kline, Brian Bond, Henry Quesada, Ben Richardson, Khris Beagley

SUSTAINABLE BIOMATERIALS

VirginiaTech Invent the Future

- Thank you!

SUSTAINABLE BIOMATERIALS

VirginiaTech Invent the Future

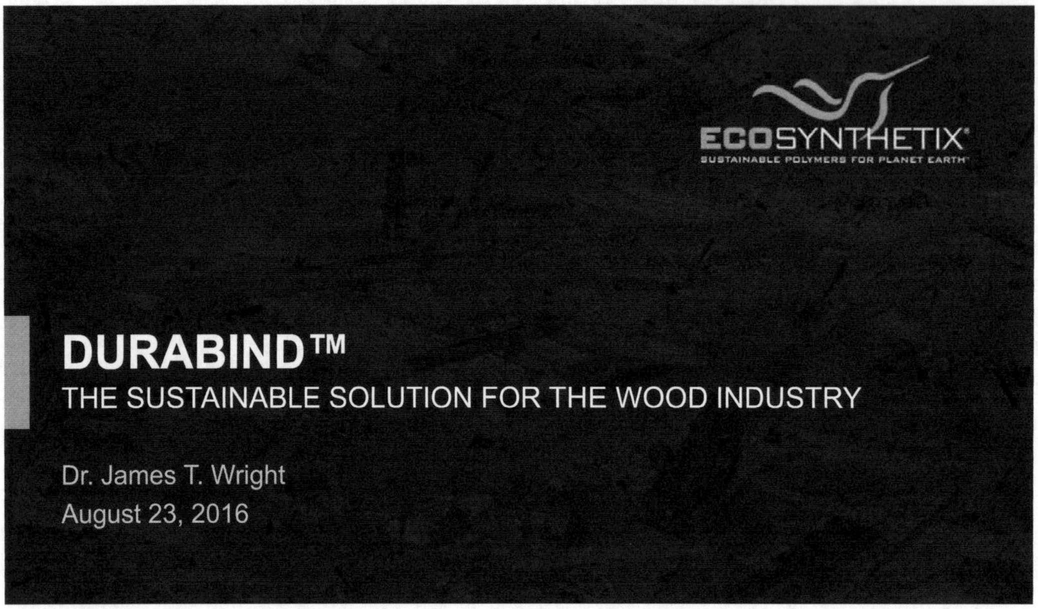

WHO WE ARE

Enhancing Nature's Best	Adding functionality - tailored to overcome the challenges of fossil fuel based materials
Viable and practical	Compatible with incumbent technologies, economically positive
Global Reach	Manufacturing, distribution, customer support in over 20 countries
Experienced Team	Global and local experienced industry team

COLLABORATIVE GLOBAL PARTNERSHIPS
Universities
Fundamental Research
EcoSynthetix
Applied Research
Centre of Innovation
Burlington, Ontario
In-house scientists, engineers and technicians
Suppliers & Customers
Open Innovation
Strategic partnerships with leading suppliers & customers
Technology
Products
Applications
LEVERAGING RELATIONSHIPS TO ACCELERATE DEVELOPMENT OF GREEN TECHNOLOGY
ECOSYNTHETIX INC. 3

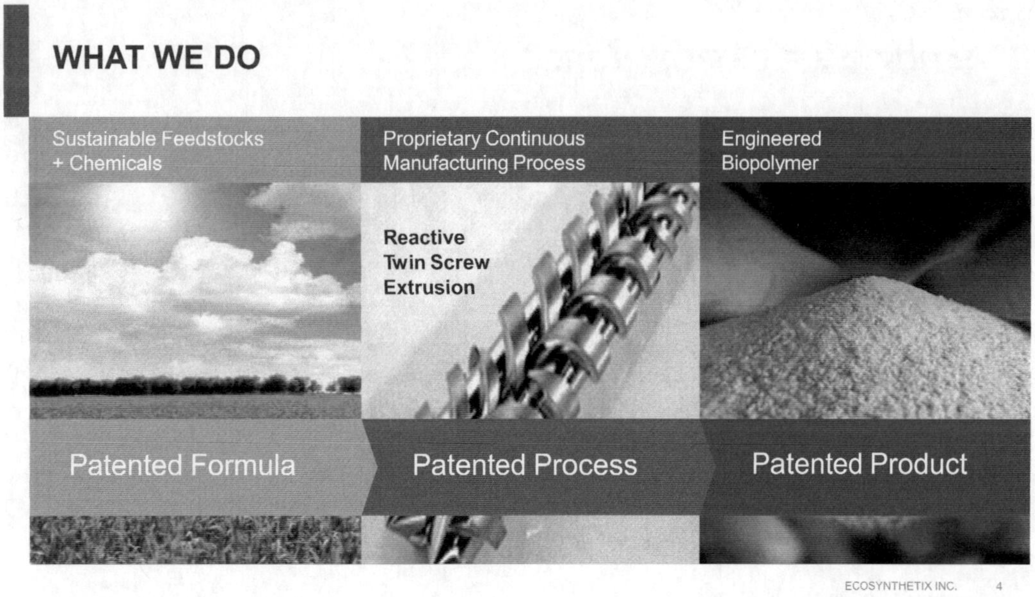
WHAT WE DO
Sustainable Feedstocks + Chemicals
Proprietary Continuous Manufacturing Process
Engineered Biopolymer
Reactive Twin Screw Extrusion
Patented Formula
Patented Process
Patented Product
ECOSYNTHETIX INC. 4

OUR ENGINEERED BIOPOLYMERS

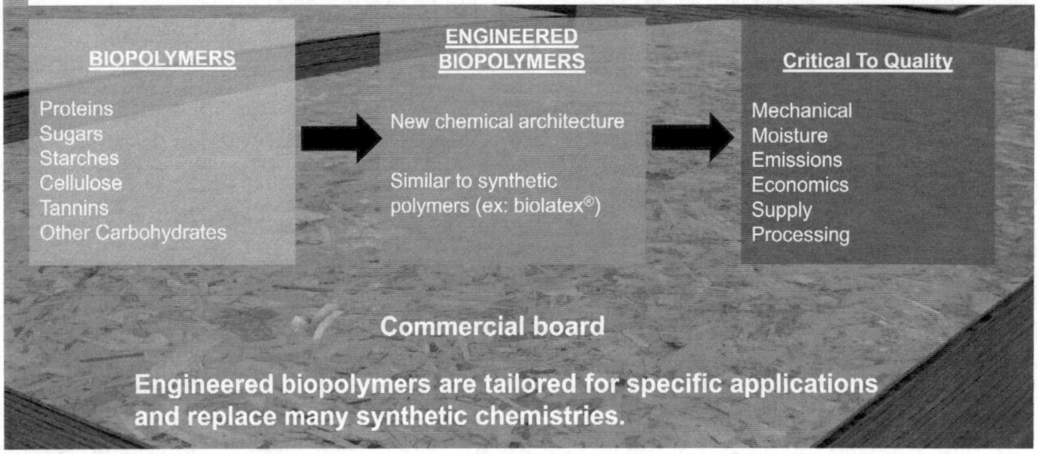

ECOSYNTHETIX INC. 5

NANO-SCALE BIOPOLYMERS

ECOSYNTHETIX INC. 6

A COMPELLING VALUE PROPOSITION

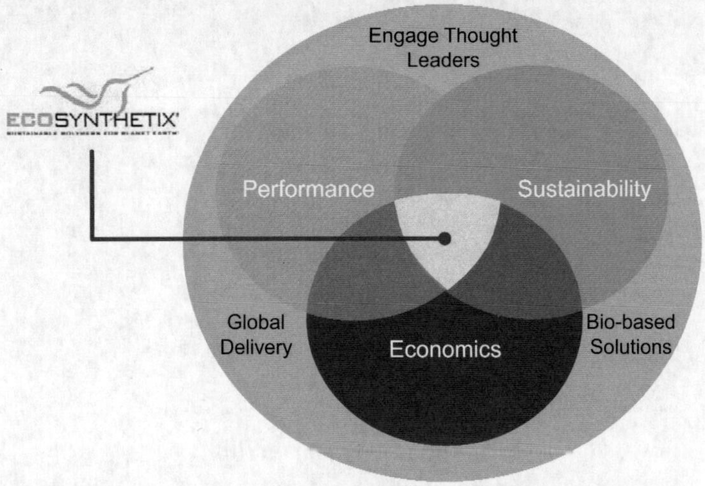

NEWS RELEASE

EPA Issues Final Rule to Protect the Public from Exposure to Formaldehyde

JULY 27, 2016 – WASHINGTON - The U.S. Environmental Protection Agency (EPA) today moves to **reduce exposure to formaldehyde vapors** from certain wood products produced domestically or imported into the United States.

One year after the rule is published, **composite wood products that are sold, supplied, offered for sale, manufactured, or imported** in the United States will need to be labeled as TSCA Title VI compliant. These products include: **hardwood plywood, medium-density fiberboard, particleboard** as well as household and other finished goods containing these products.

"**Wal-Mart Stores Inc.** is asking suppliers to **remove formaldehyde**…from their products, part of an effort to eliminate controversial chemicals from household goods.

Bloomberg News
July 21, 2016

ECOSYNTHETIX INC. 9

THE DURABIND™ SYSTEM

DURABIND BIOURETHANE™ CHEMISTRY
Tailored emulsion of engineered Biopolymer and Cross-linker
Targeted to be at least cost neutral with UF System
Lower cost versus pMDI

DURABIND PROCESS ENHANCEMENTS
Application technology: delivery systems, resin distribution manifolds
Engineering Support : from trial to commercialization
Process stability: density profile, resin distribution
Improved Performance: Reduction in Blow line contamination
Improved Machinability: Reduction in tool wear

DURABIND RENEW TECHNOLOGY
Enabling technology to reuse "squeeze" and other excess process water
Enabling scavenger to allow recycling of wood from any binder source

CHEMISTRY, PROCESS ENHANCEMENTS, RENEW

ECOSYNTHETIX INC. 10

DURABIND™ VALUE DRIVERS TO ENABLE EASE-OF-USE AND IMPROVED ECONOMICS

DuraBind Value-added Components:

Comparable product performance

Reduced maintenance and downtime

Improved machinability / tool wear

Wood chip / resin pot life

Reduction of release agents

Waste water reuse (MDF applications)

ECOSYNTHETIX INC. 11

MEETING THE STANDARD: INDUSTRIAL STRENGTH

Bond Strength

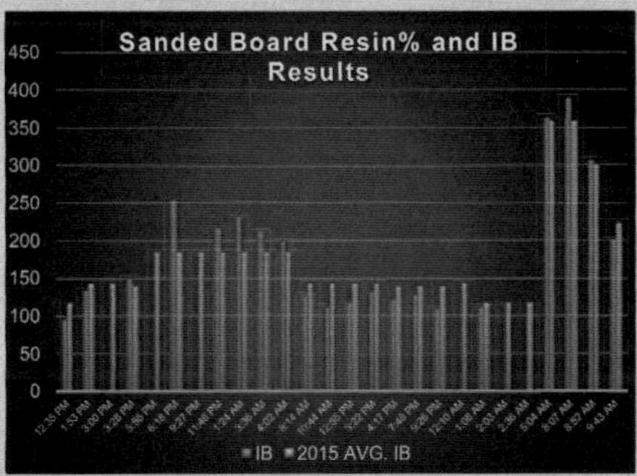

Flex Strength

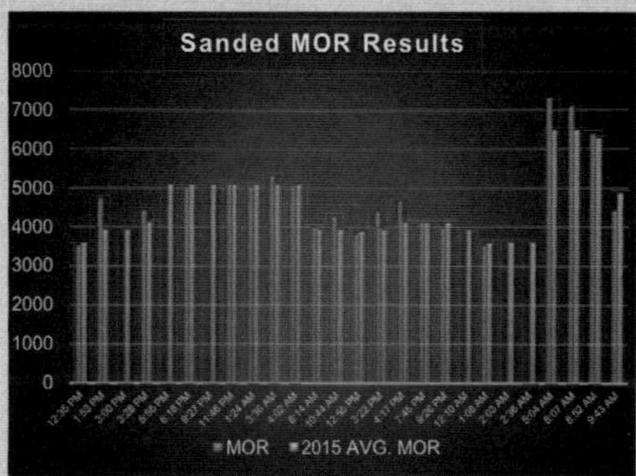

BOARDS MEET SPEC IN INDUSTRIAL APPLICATIONS

ECOSYNTHETIX INC. 12

DURABIND^{TM} BENEFITS – MDF INDUSTRIAL APPLICATION

CONFIDENT IN COMMERCIAL BOARD PERFORMANCE

Sample (8mm flooring MDF)	Control (2.7% pMDI)	DuraBind™ 25 Bio-urethane (2.7% loading)
Density	846	834
MOR (N/mm2)	42.5	41.7
IB (N/mm2)	2.5	2.3
2h TS	12.4	10.7
24h TS	22.3	22.3
2h WA	19.2	18.0
24h WA	34.3	35.2

ECOSYNTHETIX INC. 13

DURABIND^{TM} BENEFITS – THIN MDF INDUSTRIAL APPLICATION

QUALITY MAINTAINED WITH REDUCED LOADING

3 mm MDF		Control	Phase 1	Phase 2	Phase 3	Phase 4
UF	%	10				
pMDi	%	-	2.4	2.0	1.75	2.0
DuraBind	%	-	0.67	0.67	0.67	0.67
Total system loading	%	-	3.07	2.67	2.42	2.67
wax:	%	1.5	1.5	1.5	1.5	1.5
Press speed	mm/s	1350	1150	1150	1150	1275
Raw thickness:	mm	3.2	3.2	3.24	3.28	3.25
Density	Kg/m3	861	829	843	833	833
IB:	N/mm2	1.58	1.92	1.74	1.85	1.76
Surface Soundness	N/mm2	1.64	1.73	1.45	1.91	1.50
Swelling 24 h	%	33.3	23.9	25.6	33.4	24.6
Absorption 24 h.	%	51.7	42.8	48	64.7	43.1

ECOSYNTHETIX INC. 14

DURABIND™ BENEFITS – INDUSTRIAL PB TRIAL RESULTS

DEMOSTRATED IMPROVEMENT IN WATER TOLERANCE

	CONTROL	DURABIND
I.B. (N/mm²)	0.55	0.53
M.O.R. (N/mm²)	11.0	10.5
M.O.E. (N/mm²)	2280	2320
2h T.S. (%)	4.3	2.6
24h T.S. (%)	23.5	20.5
Formaldehyde content (mg/100g)	6.5	2.2
Moisture content (%)	6.3	5.1
Surface soundness (pull) (N/mm²)	1.12	1.06
Density (kg/m³)	675	675

DURABIND™ BENEFITS – INDUSTRIAL OSB PRODUCTION RESULTS

DEMOSTRATED Results while producing over 70,000 Cubic Meters

11 MM OSB PANEL	Control (2.6% pMDI)	DuraBind™ 25 Bio-urethane
Density	620	620
MOR (N/mm2)	25.4	27.8
M.O.E. (N/mm2)	4460	4570
24h T.S. (%)	22	25
I.B. (N/mm2)	0.37	0.41
IB after Boiling (N/mm2)	0.11	0.13

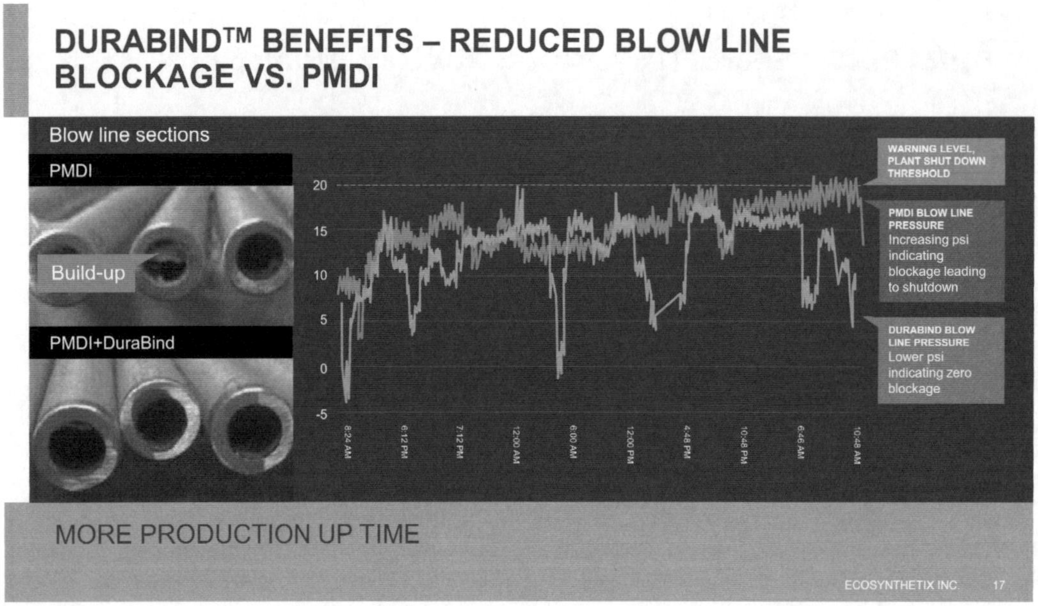
DURABIND™ BENEFITS – REDUCED BLOW LINE BLOCKAGE VS. PMDI
Blow line sections
PMDI
Build-up
PMDI+DuraBind
WARNING LEVEL, PLANT SHUT DOWN THRESHOLD
PMDI BLOW LINE PRESSURE
Increasing psi indicating blockage leading to shutdown
DURABIND BLOW LINE PRESSURE
Lower psi indicating zero blockage
20
15
10
5
0
-5
8:24 AM
6:12 PM
7:12 PM
12:00 AM
6:00 AM
12:00 PM
4:48 PM
10:48 PM
6:46 AM
10:48 AM
MORE PRODUCTION UP TIME
ECOSYNTHETIX INC. 17

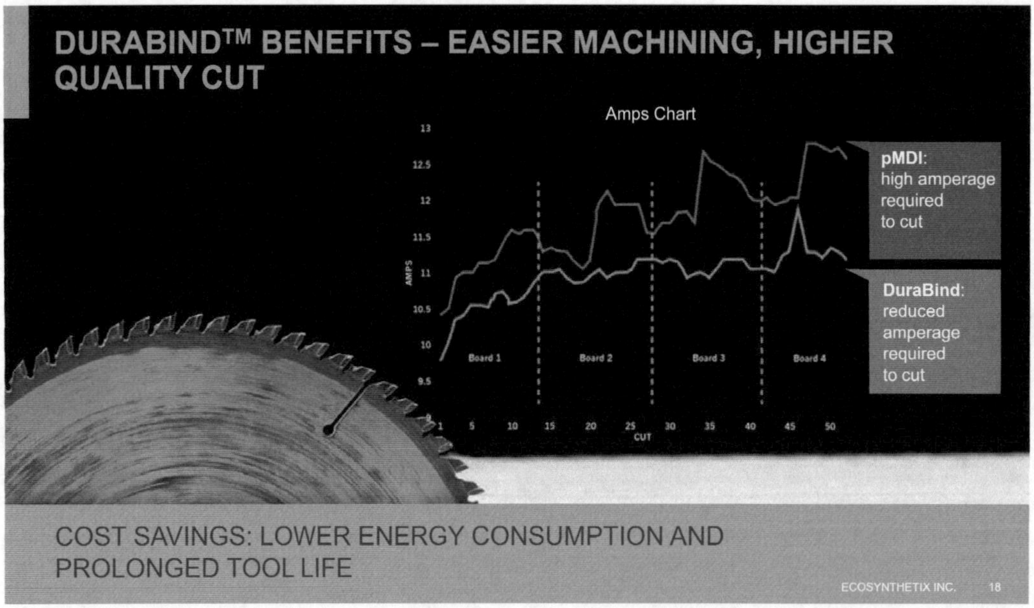
DURABIND™ BENEFITS – EASIER MACHINING, HIGHER QUALITY CUT
Amps Chart
AMPS
CUT
Board 1
Board 2
Board 3
Board 4
pMDI:
high amperage required to cut
DuraBind:
reduced amperage required to cut
COST SAVINGS: LOWER ENERGY CONSUMPTION AND PROLONGED TOOL LIFE
ECOSYNTHETIX INC. 18

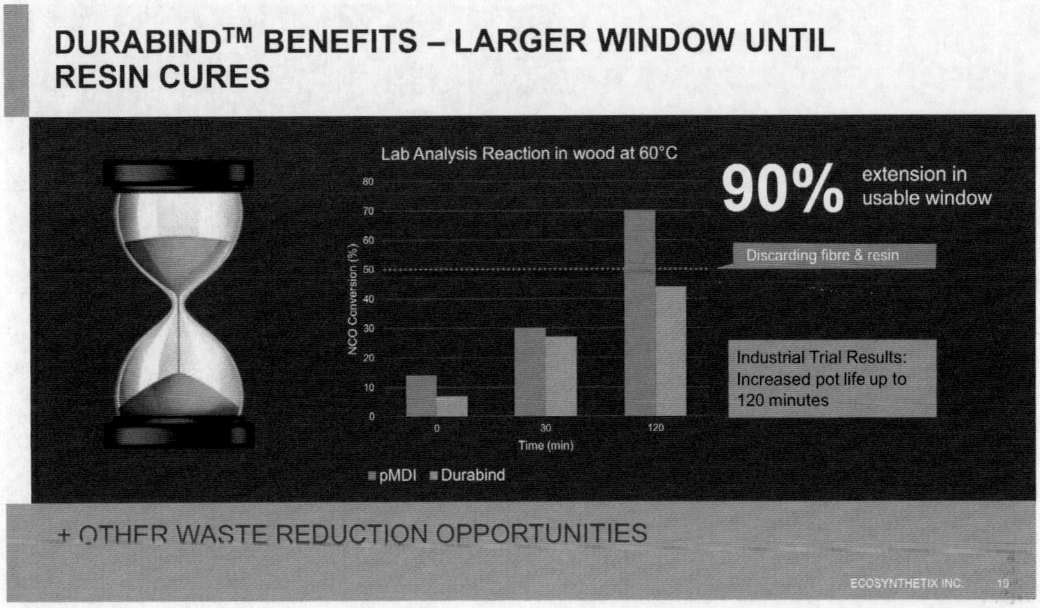
DURABIND™ BENEFITS – LARGER WINDOW UNTIL RESIN CURES
Lab Analysis Reaction in wood at 60°C
NCO Conversion (%)
80
70
60
50
40
30
20
10
0
0
30
120
Time (min)
pMDI
Durabind
90% extension in usable window
Discarding fibre & resin
Industrial Trial Results: Increased pot life up to 120 minutes
+ OTHER WASTE REDUCTION OPPORTUNITIES
ECOSYNTHETIX INC.
19

DURABIND™ BENEFITS – IMPROVED WASTE WATER MANAGEMENT
Waste water from MDF refiner process
Waster water recycled on site to disperse DuraBind
Disposal Cost
up to $20,000
per week
$0
per week
+ OTHER WASTE REDUCTION OPPORTUNITIES
ECOSYNTHETIX INC.
20

DURABIND™ BENEFITS – ELIMINATE BOARD-TO-PRESS STICKING

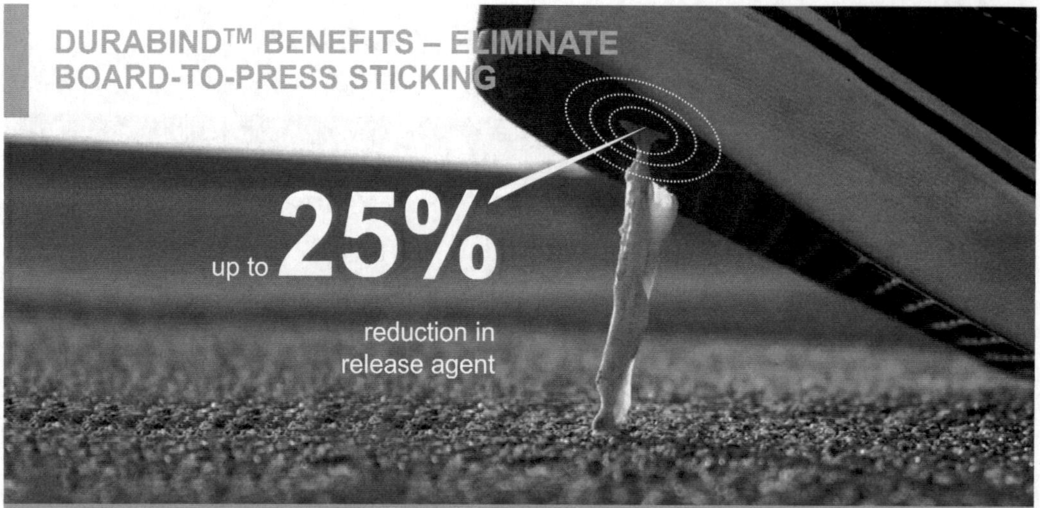

REDUCED RELEASE AGENT COSTS

ECOSYNTHETIX INC. 21

DURABIND™ WELL-POSITIONED FOR BROAD ADOPTION

PERFORMANCE

+

VALUE

LONGER PRODUCTION UP-TIME
Reduced blow-line blockage

LOWER PRODUCTION COSTS
Longer tool life
Lower use of costly processing aids

LESS WASTE
Larger processing window

BETTER QUALITY, HIGHER VALUE BOARDS
Market share & margin for our customers

ECOSYNTHETIX INC. 22

ECOSYNTHETIX®
SUSTAINABLE POLYMERS FOR PLANET EARTH™

EcoSynthetix Inc.

Question & Answer

EVERGREEN ENGINEERING
Engineering and Construction Services

CLT

EVERGREEN ENGINEERING
Engineering and Construction Services

Overview

A look at the addition of a CLT production facility for American lumber producers

EVERGREEN ENGINEERING
Engineering and Construction Services

Outline

- Market Outlook
- The Basics
- Production Overview
- Configuration and Cost
- Scalability
- CapEx/OpEx
- Broad Financial Strokes

EVERGREEN ENGINEERING
Engineering and Construction Services

Market Outlook

While today, we are talking about an early-stage, fledgling industry, the demand is projected to rise substantially.

U.S. production is currently just under 34,000 m^3.

Production Capacity

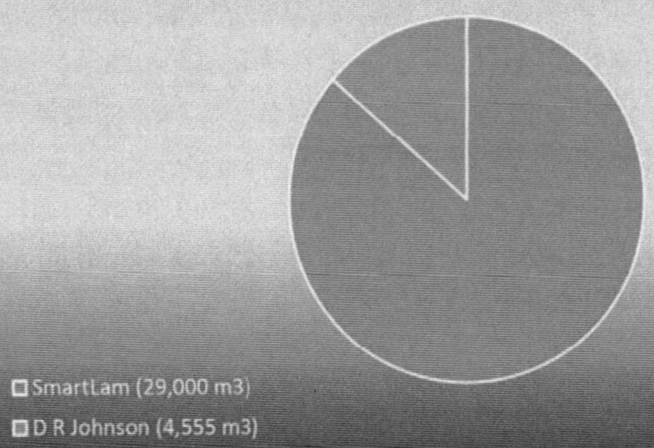

EVERGREEN ENGINEERING
Engineering and Construction Services

Market Outlook

By 2020 according to several sources, demand could top 300,000 m^3.

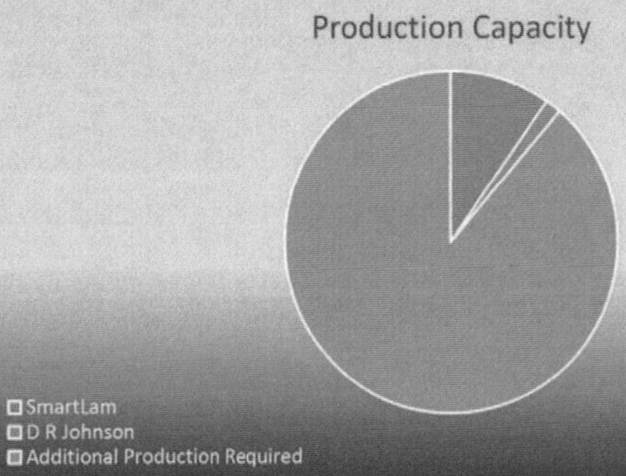

EVERGREEN ENGINEERING
Engineering and Construction Services

Market Outlook

- There are still less than 20 CLT structures completed in North America.
  - Many of these structures were constructed of panels not produced in the U.S.
  - So far the U.S. has two mills that have come to the CLT table via different but important synergies.
    - Smartlam was the first producer of CLT selling panels to the oil and gas industry as an excellent solution to their platform and heavy equipment needs as industrial matting.
    - D R Johnson, having already established a position in the architectural market with GluLam beams, had some of the equipment in place to make CLT for building.

EVERGREEN ENGINEERING

Engineering and Construction Services

Market Outlook

- Projected demand by 2020 represents a ten-fold increase over current production.
- These numbers indicate a potential for 3 to 5 new CLT production facilities, even after existing producers reach full capacity.
- Following the European experience, projected demand triples between 2020 and 2025.

EVERGREEN ENGINEERING

Engineering and Construction Services

The Basics

What You Need

- Wood
- Adhesive
- Pressure

What you do

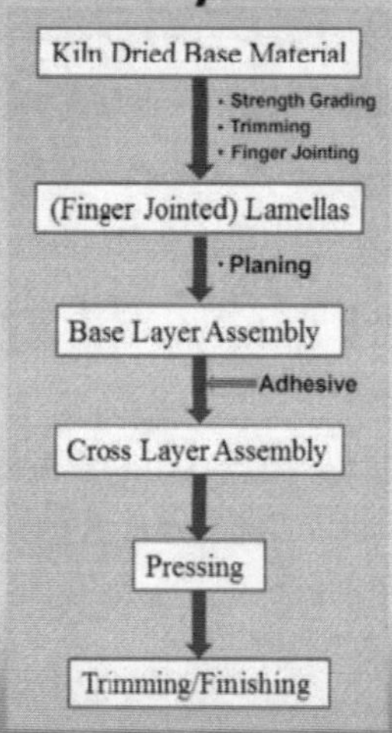

EVERGREEN ENGINEERING
Engineering and Construction Services

Production Overview

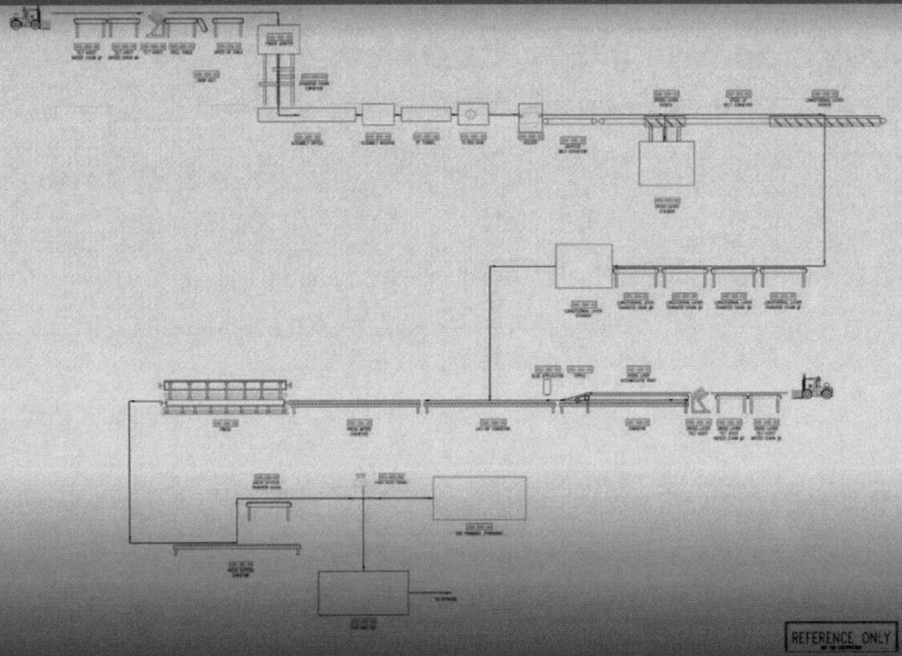

EVERGREEN ENGINEERING
Engineering and Construction Services

Configuration & Cost

In a manual panel assembly, production panels are placed on the reversing layup conveyor by hand. The panel is then fed via the conveyor under the glue applicator, then reversed and the second transverse layer is placed and that layer is glued. This process is repeated until the desired panel thickness is achieved.

This process generally takes an hour per panel, and the glue set time is the primary production constraint.

This method requires 7 employees

EVERGREEN ENGINEERING
Engineering and Construction Services

Configuration & Cost

In an automated lay-up design, assembly time is reduced from one hour to 10 minutes.

Because the assembly time is reduced, glue workability is less of an issue and faster setting glues can be used.

Press times can be reduced and in Europe, press times of 30 minutes have been achieved.

Automated lay-up has a operator requirement of 5, reducing headcount by 2.

EVERGREEN ENGINEERING
Engineering and Construction Services

Configuration & Cost

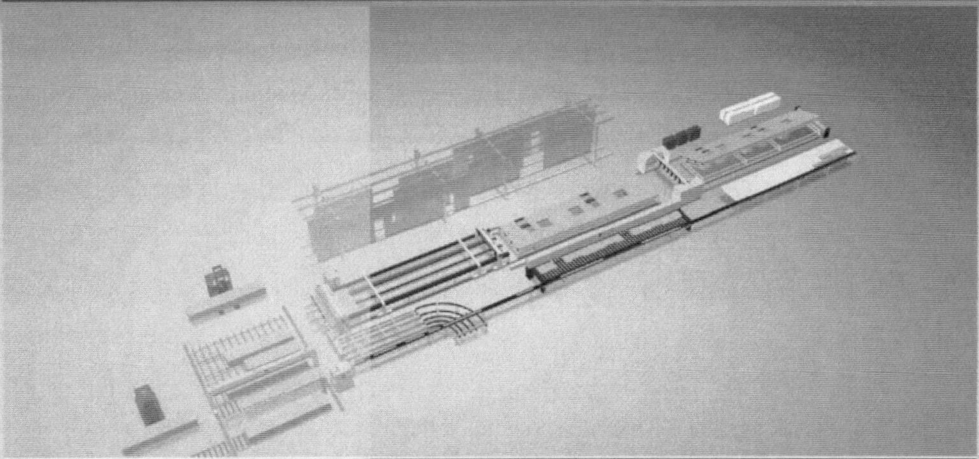

WEINIG

Spezialized Cross Laminated Timber

EVERGREEN ENGINEERING
Engineering and Construction Services

Configuration & Cost

The capital cost projected at this time for a manual system capable of 30,000m^3.

Process System	Total Direct Cost	Indirect Cost	Total Capital Cost
Manual	$9,851,763	$5,335,236	$15,244,916
Automated	$11,643,449	$6,166,101	$17,878,000

EVERGREEN ENGINEERING
Engineering and Construction Services

Scalability

Building to meet market demand now and with an eye to the future will require designing for scalability.

- Engineer for future automation.
- Identify potential bottlenecks.
- Consider creative solutions.
    - In some mills, European producers are making 20'x60' panels and cutting them into two, and for the U.S. market, possibly four 10'x30' panels.

EVERGREEN ENGINEERING
Engineering and Construction Services

Broad Financial Strokes

- Profitability studies thus far are assuming a static average price per panel.
- Real-world costs per panel will fluctuate with markets.
- Opportunities for cost reduction lie in automation, adhesive cost, and lumber usage via grade recovery (defect elimination).

EVERGREEN ENGINEERING
Engineering and Construction Services

Discussion

EVERGREEN ENGINEERING

Engineering and Construction Services

Who We Are

Since 1985, Evergreen Engineering has been committed to providing innovative industrial engineering, design, consulting, and construction solutions for a variety of industries. Our ability to understand the needs of our clients goes hand-in-hand with our knowledge of industry processes. We have a long list of satisfied clients because we have created a dynamic team with the technical depth to facilitate a diverse range of successful projects.

EVERGREEN ENGINEERING

Engineering and Construction Services

Thank You

Overview

1. What do we mean by Mass Timber Construction?
2. Constructability
3. Drivers
4. Current roadblocks

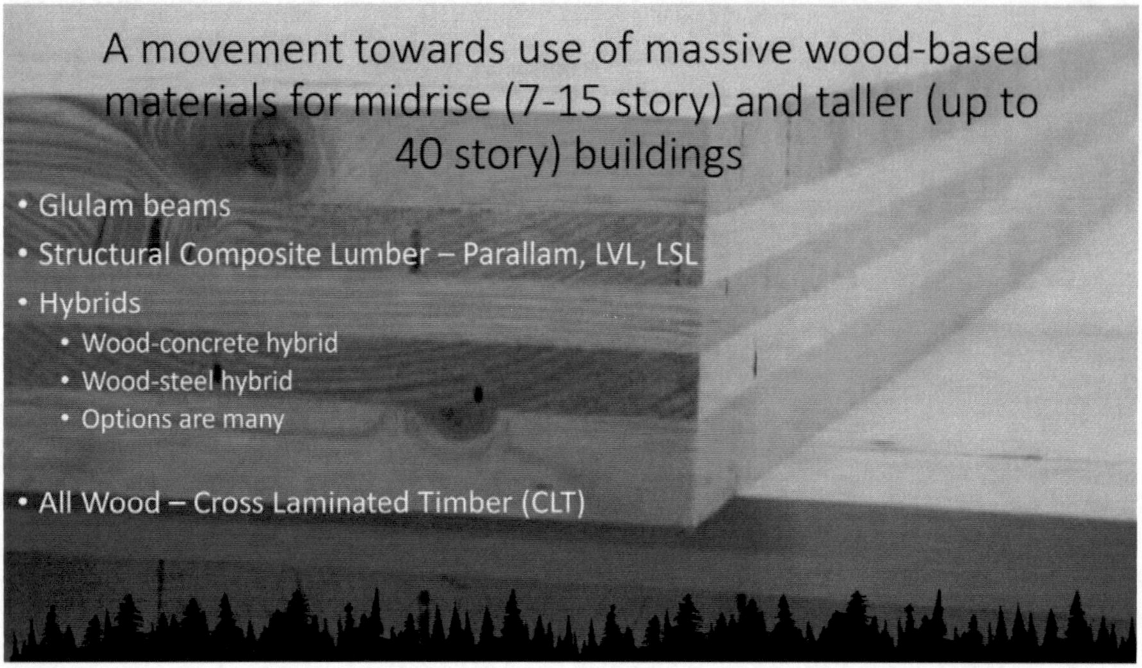
A movement towards use of massive wood-based materials for midrise (7-15 story) and taller (up to 40 story) buildings
• Glulam beams
• Structural Composite Lumber – Parallam, LVL, LSL
• Hybrids
• Wood-concrete hybrid
• Wood-steel hybrid
• Options are many
• All Wood – Cross Laminated Timber (CLT)

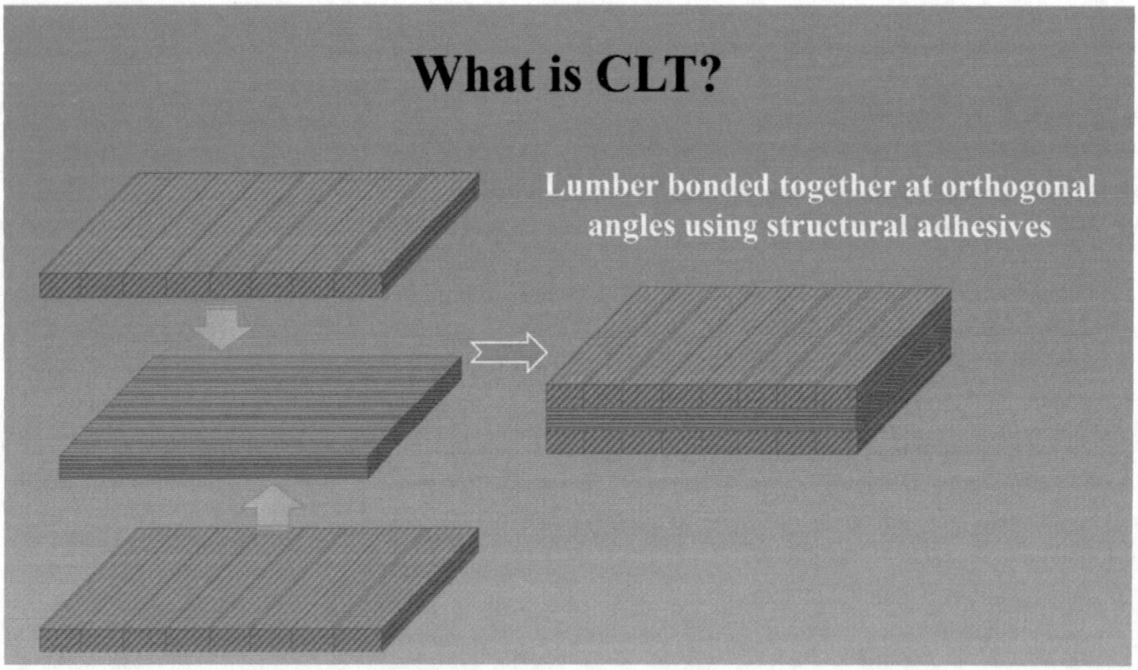
What is CLT?
Lumber bonded together at orthogonal angles using structural adhesives

Pictures courtesy of linshapespace.com, construction.com

Pictures courtesy of woodskyscapers.com

Lumber Specs
Any softwood lumber species recognized by ALSC or CLSAB may be used – SG > 0.35 (ANSI/APA PRG_320)
Common Specifications:
8-10 ft. length
5/8 in. to 2 in. thick
12% Moisture content
CLT Handbook
CROSS-LAMINATED TIMBER
US EDITION
Standard for Performance-Rated Cross-Laminated Timber
ANSI
APA

Laminate Architecture

Major strength direction: Minimal grade of 1200f-1.2E MSR or visual grade No. 2

Minor strength direction: Minimal grade -- visual grade No. 3

Structural adhesive must meet the requirements in the AITC 405 standard

More information can be found in the PRG_320 2012 and CLT handbook

Source: CLT Handbook and ANSI/APA PRG_320 2012

Panel Dimensions

Maximum width of 8-10 ft. per shipping requirements

Maximum length up to 40 ft. due to shipping requirements

Maximum thickness of 20 in.
Design governed by deflection and/or vibration
(depth varies -- 4" to 10" typically)

Courtesy of fireengineering.com

CLT Handbook and ANSI/APA PRG_320 2012

CLT Examples
Courtesy of architectureanddesign.com.
Courtesy of yahoo finance
Courtesy of europeanwood.org
Forte
10 stories tall
Melbourne, Australia
The tree
62 luxury apartments
Bergen, Norway
Stadthous
9 stories of residential suites
Located in London

9-Story CLT Building
https://www.youtube.com/watch?v=tJHeLyFASEI
Stadthaus, Murray Grove
London
Architect: Waugh Thistleton Architects Ltd.
Structural Engineer: Techniker
Timber by KLH UK Ltd.
Client: Telford Homes / Metropolitan Housing Trust
Photographs: Will Pryce

Week 0
Week 1
Week 2
Week 3
Week 4
Week 5
Week 5

Week 6
Week 7
Week 8
Week 9
Done!

CLT in construction

- Typical layout
- Common connections
- Construction methods

Example Layout

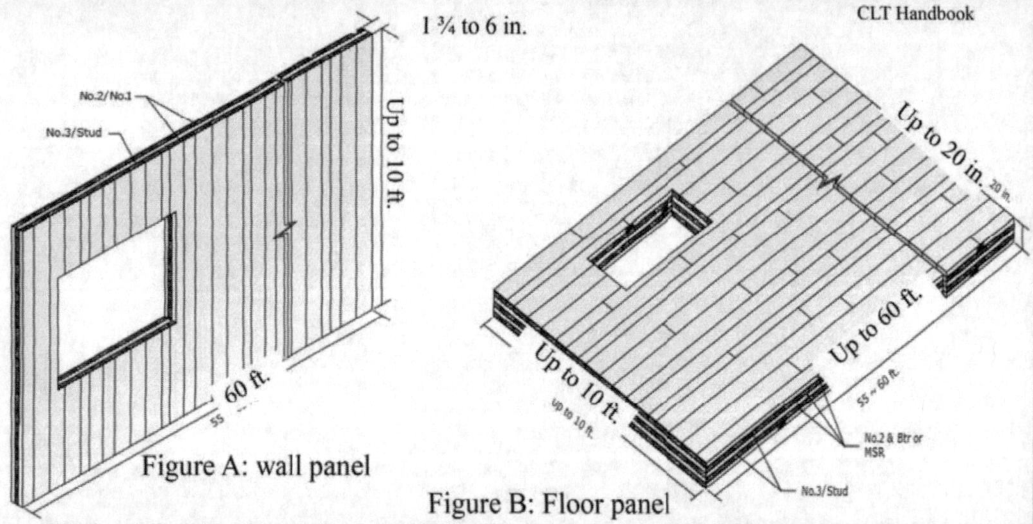

Figure A: wall panel

Figure B: Floor panel

CLT Connections

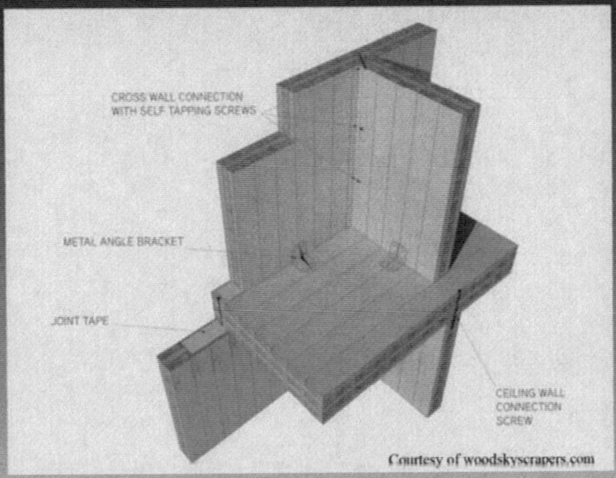

Connection Systems commonly used:

Long self-tapping screws – Used in panel to panel i.e. floor to floor or floor to wall.[1]

Bolts and Dowels – Used commonly for lateral loads.

Innovative type fasteners – Self-tapping screw and glued-in rods.

(1) CLT Handbook 2013

On-site construction

- Similar to that of concrete panels
- Less time to construct and less wasted materials on job site

Construction Apprehensions

The storage of CLT panels is important – avoid standing water

Lifting methods have been specifically adapted to accommodate CLT panels

Transportation is one of the key issues driving up the cost of the panels due to limited manufacturers at this time

Benefits compared to concrete

- 25% faster to construct
- 90% less construction traffic
- 75% less deck labor
- Significantly less site noise

Construction of Forte in Melbourne, Australia
https://www.youtube.com/watch?v=cqXygHyU5ws

Source: Bernhard Gafner, "Mass Timber Building Systems: Understanding the Options"
http://www.woodworks.org/wp-content/uploads/NE-WSF-150916-GAFNER-Mass-Timber-Building-Systems.pdf

0:36 / 2:13

1:00 / 2:13

FORTE
1:38 / 2:13

2:02 / 2:13

Environmental Impacts

Reduced Carbon footprint

Sustainable Forests

Carbon Footprint

¼ the carbon footprint of steel and concrete structures[2]

Steel and concrete add up to 8% of greenhouse gases[3]

1 m^3 of wood stores 1 tonne of CO_2 [3]

(2) CLT: the future of building (3) Michael Green Architecture

Promotes Sustainable Practices

- Sustainable timber harvesting, which is already practiced in North America, goes hand-in-hand with Mass Timber Construction
- Small diameter timber can be used
- Contributes to healthy forests

(2) CLT: the future of building

When designing panels, must keep in mind…

In-plane and out-of-plane bending stresses, shear stress and stiffness

Durability in adverse environments

Short and long term loading (creep)

Fire and seismic protection

Vibration performance of floors and sound insulation

Charring performance

a) Unprotected CLT before test b) Unexposed surface during test c) Protected CLT after test

CLT fire-resistance wall tests conducted at NRCC in Ottawa (Canada) 2013 CLT handbook

CLT is incorporated into the 2015 edition of the International Building Code (IBC), with an unprotected fire resistance rating of up to two hours[1]

According to the National Building Code published by American Insurance Association a cast-in-place concrete 5 1/2-inch wall has a fire resist rating of 2 hours[5]

(5) Fire engineering (6) concreteconstruction.net

Seismic Concerns

Seismic ratings for CLT are not yet incorporated in the *Quantification of Building Seismic Performance Factors FEMA P695 / June 2009*

Research is ongoing on the CLT systems and individual panels

Seven-story CLT system tested on a shake table at E-defense labs in Miki, Japan; a joint effort from the SOFIE Project and CLT test assembly at FPInnovation[1]

[1] CLT Handbook 2013

2013 CLT Handbook

Evaluation of connection performance
by Dan Dolan at WSU

Thank you!

Questions?

AUTHOR INDEX